Superoxide Dismutase

Volume I

Editor

Larry W. Oberley, Ph.D.
Associate Professor of Radiobiology
Radiation Research Laboratory
University of Iowa
Iowa City, Iowa

CRC Press, Inc.
Boca Raton, Florida

Library of Congress Cataloging in Publication Data

Main entry under title:

Superoxide dismutase.

Includes bibliographies and indexes.
1. Superoxide dismutase. 2. Active oxygen.
I. Oberley, Larry W.
QP603.S94S97 574.19′258 81-21683
ISBN 0-8493-6240-7 (v. 1) AACR2
ISBN 0-8493-6241-5 (v. 2)

This book represents information obtained from authentic and highly regarded sources. Reprinted material is quoted with permission, and sources are indicated. A wide variety of references are listed. Every reasonable effort has been made to give reliable data and information, but the author and the publisher cannot assume responsibility for the validity of all materials or for the consequences of their use.

Direct all inquiries to CRC Press, Inc., 2000 Corporate Blvd., N.W., Boca Raton, Florida, 33431.

International Standard Book Number 0-8493-6240-7 (Volume I)
International Standard Book Number 0-8793-6241-5 (Volume II)

Library of Congress Card Number 81-21683
Printed in the United States

PREFACE

This book is about oxygen radicals and the living cell's defense mechanisms against these species. It describes the production of these radicals, their chemistry, the enzyme superoxide dismutase, its structure and function, and the role of these entities in biological systems. The significance of oxygen radicals in various pathological conditions will also be discussed.

The work described herein is mainly due to the vision of one man. As modern physics is based on the work of Bohr and molecular biology is built on the work of Watson and Crick, so the study of oxygen radicals is built on the work of Dr. Irwin Fridovich (realizing, of course, that he had excellent help, especially from Dr. Joe McCord). Others have claimed to have discovered superoxide dismutase before Dr. Fridovich. However, we should remember that Dr. Fridovich's main contribution is not that of being the first to purify the enzyme, but rather that he was the first to explain its function and significance. More than any other person, he has guided subsequent work. This book is a tribute to his genius.

I hope these are the first two of several volumes. The next volumes will deal with superoxide dismutase and disease. The last chapter in these two volumes is a preview of this subject area.

This book is organized as follows. In the first volume, Dr. Fridovich discusses the history of superoxide dismutase in Chapter 1. In Chapter 2, Dr. Steinman elucidates what is known about the structure of this enzyme. Dr. Fridovich describes the assay of this enzyme in Chapter 3. Chapters 4 and 5 are devoted to the function of this enzyme-protection against the destructive effects of superoxide radicals. In Chapter 4, Dr. Fridovich explains the importance of SOD in defense against oxygen toxicity in prokaryotic cells. Dr. Halliwell talks about oxygen toxicity in plant tissues in Chapter 5. It should be remembered when reading these chapters that oxygen toxicity is also seen in higher organisms, like man. The next three chapters are devoted to production of superoxide radical, the substrate for superoxide dismutase, in chemical and biological systems. In Chapter 6, Drs. Allen and Bielski discuss the production of superoxide radicals in aqueous solutions. In Volume II, Drs. Gabig and Babior explain the role of oxygen radicals in killing of microbes in Chapter 1. At the subcellular level, Drs. Boveris and Cadenas discuss the production of superoxide and hydrogen peroxide by mitochondria in Chapter 2. The next two chapters are devoted to the methods for detecting oxygen radicals. In Chapter 3, Dr. Bors outlines the various methods used for measuring superoxide and hydroxyl radicals. In Chapter 4, Dr. Buettner explains one of the newly discovered and most useful techniques for detecting oxygen radicals — spin trapping. Chapter 5 is devoted to the organic chemistry of superoxide anion radical. In this chapter, Dr. Frimer is mainly concerned about the chemistry of superoxide radical in nonaqueous media. The last chapter, written by myself, deals with the role of superoxide dismutase in cancer. This last chapter is built on all the others, because knowledge of the basic chemistry and biology of superoxide dismutase is necessary to understand the role of this enzyme in disease.

It is always dangerous to speculate about the significance of any research, especially when one is involved in that area and is thus somewhat biased. Nevertheless, with all the attention given to other areas of biochemistry, especially recombinant DNA, I feel some speculation is in order. I think the subject of oxygen radicals in disease will be the most important field of research in medical sciences for some time to come, eclipsing some of the present very popular subject areas. This is because any disease must have as its basis some mechanism of damage, and that mechanism will most likely be shown to be the very reactive oxygen radicals. Thus, oxygen radicals have already been implicated in immune diseases, cancer, aging, diabetes, and muscular dystrophy.

Oxygen radicals may not be the initial cause of those diseases, but they may cause the ultimate cell damage. I predict that knowledge of this area will be essential to future medical treatment.

THE EDITOR

Larry W. Oberley, Ph.D., is Associate Professor of Radiology in the College of Medicine of the University of Iowa in Iowa City, Iowa. Dr. Oberley was graduated from Northwestern University, Evanston, Illinois with a B.A. in Physics in 1968. He received an M.S. in physics from the University of Iowa in 1980. During the next five years, Dr. Oberley was employed as a research associate in the Division of Nuclear Medicine, Department of Radiology, at the University of Iowa. It was during this period that Dr. Oberley developed his interest in biochemistry and medicine. In 1974, Dr. Oberley received his Ph.D. in physics from the University of Iowa. Dr. Oberley was appointed Instructor in the Radiation Research Laboratory of the Department of Radiology at the University of Iowa in 1975, Assistant Professor in 1978, and Associate Professor in 1980.

Dr. Oberley is a member of several societies including Sigma Xi, the New York Academy of Sciences, the Society of Nuclear Medicine, the American Society for Photobiology, the Radiation Research Society, and the American Association for the Advancement of Science.

Dr. Oberley has published over 30 articles and has given presentations at more than 20 scientific meetings. His current research interests are the role of free radicals in cancer, radiation, and aging.

CONTRIBUTORS

Augustine O. Allen, Ph.D.
Consultant
Brookhaven National Laboratory
Upton, New York

Bernard M. Babior, M.D.
Professor of Medicine
Tufts University School of Medicine
New England Medical Center Hospital
Boston, Massachusetts

Benon H. J. Bielski, Ph.D.
Chemist
Brookhaven National Laboratory
Upton, New York

Wolf Bors, D. Sc.
Research Scientist
Institut für Biologie
Gesellschaft für Strahlen- und Umweltforschung
Neuherberg
Federal Republic of Germany

Alberto Boveris, Ph.D.
Associate Professor
Department of Biochemistry
University of Rio de Janeiro
Rio de Janeiro, Brazil

Garry R. Buettner, Ph.D.
Assistant Professor of Chemistry
Wabash College
Crawfordsville, Indiana

Enrique Cadenas, M.D., Ph.D.
Research Associate
Johnson Research Foundation
School of Medicine
University of Pennsylvania
Philadelphia, Pennsylvania

Irwin Fridovich, Ph.D.
James B. Duke Professor
Department of Biochemistry
Duke University Medical Center
Durham, North Carolina

Aryeh A. Frimer, Ph.D.
Senior Lecturer
Department of Chemistry
Bar Ilan University
Ramat Gan
Israel

Theodore G. Gabig, M.D.
Chief, Division of Hematology
Veterans Administration Hospital
Ann Arbor, Michigan

Barry Halliwell, D.Phil.
University Lecturer in Biochemistry
Kings College
University of London
London
England

Christa Michel, Ph.D.
Research Assistant
Institut für Biologie
Gesellschaft für Strahlen- und Umweltforschung
Neuherberg
Federal Republic of Germany

Larry W. Oberley, Ph.D.
Associate Professor
Radiation Research Laboratory
Department of Radiology
University of Iowa
Iowa City, Iowa

Manfried Saran, Ph.D.
Research Scientist
Institut für Biologie
Gesellschaft für Strahlen- und Umweltforschung
Neuherberg
Federal Republic of Germany

Howard M. Steinman, Ph.D.
Associate Professor
Department of Biochemistry
Albert Einstein College of Medicine
Bronx, New York

ACKNOWLEDGMENTS

I would like to thank Susan Barnett and Carol Kieffer for their help in manuscript preparation. I would like to thank the contributing authors for their time and effort.

The support of the family of my childhood — Ruby, Jim, and Terry — and my present family — Kathleen, Rebecca, and Christopher — is gratefully acknowledged. Without their encouragement and patience, this volume would never have been completed.

Last, I must thank medical science for allowing me to edit and write this volume. I have been an artificial kidney patient for the last eight years. It is sobering to think that if I had been born 20 years earlier, I would not have been able to work on this book. As my life has been prolonged by the wonders of science, I hope that the lives of others will be prolonged by the fruits of the knowledge contained in this volume.

DEDICATION

To Kathleen, my partner through life,
and to Rebecca and Christopher,
who I hope will benefit
from this work someday

TABLE OF CONTENTS

Volume I

Volume II

Chapter 1

THE DISCOVERY OF SUPEROXIDE DISMUTASES: A HISTORY

Irwin Fridovich

TABLE OF CONTENTS

I. INTRODUCTION

Science has been described as an edifice of great beauty and of imposing proportions constructed by pygmies, each of whom adds a block to the growing structure and each of whom must climb up on the blocks laid down by predecessors, before being in a position to make his own contribution. The authors of this volume have been privileged to be among those able to add to the growing house of Science. It is appropriate to begin this volume with an account of how this wing of the building got started.

Enzymes, such as the superoxide dismutases (SODs) whose substrate is an unstable free radical, with but a fleeting lifetime under ordinary conditions, could hardly have been discovered by design — the more so since the very existence of the substrate radical, in biological systems, was only dimly perceived before the enzymes were discovered. In fact, appreciation of the biological significance of the superoxide radical ($O_2^{\cdot-}$) and of the functions of the SODs have developed hand in hand and depended, in large measure, upon serendipity, persistence, talented graduate students and postdoctoral fellows, and liberal granting agencies. There were really three converging paths which led independently, tortuously and yet, seemingly inexorably, to an appreciation of the SODs. The first was the sulfite oxidation trail, the second was the cytochrome *c* reduction trail, and the third was the luminescence trail. We will briefly retrace each of these and see how they led to the same place.

II. THE SULFITE OXIDATION TRAIL

Sarcosine oxidase, which oxidizes *N*-methyl glycine to formaldehyde plus glycine, was discovered at Duke University by Handler in 1941.[1] It was still under study there in 1952, when I enrolled as a graduate student in biochemistry under the tutelage of Dr. Handler. Heimberg, then a senior graduate student in the Department of Biochemistry, was preparing α-hydroxysulfonic acids as analogs of sarcosine, in the expectation that they might serve as competitive inhibitors of sarcosine oxidase. I joined him in this work and we were surprised to find that α-hydroxysulfonic acids did not inhibit the oxidation of sarcosine by crude liver preparations, but were rather themselves rapidly oxidized.

α-Hydroxysulfonic acids are merely the bisulfite adducts of aldehydes or ketones, and they can dissociate, giving up bisulfite or sulfite plus the parental carbonyl compound. We soon learned that the apparent oxidation of α-hydroxysulfonic acids, by liver brei, was really an oxidation of sulfite to sulfate. This catalytic oxidation of sulfite was of great interest because sulfite was known to be an intermediate in the catabolism of cysteine and because nothing was then known about the biological oxidation of sulfite to the stable excretory form, sulfate. We gladly gave up on sarcosine oxidase and began to explore sulfite oxidation in liver. A partial purification of the responsible enzyme was achieved.[2] Heimberg wrote his thesis and went on to other endeavors in other places and I was left with the embryonic sulfite oxidase problem. The enzyme was found to be localized to the mitochondria and it was amenable to further purification.[3]

III. BIFURCATION IN THE SULFITE OXIDASE TRAIL

Sulfite has a marked tendency to oxidize by a free radical chain mechanism. This fact was in the literature,[4] but our ignorance of it forced us to learn it the hard way. Thus, we noted two distinct types of sulfite oxidation catalyzed by liver extracts. The first of these was a genuine enzymatic oxidation of one sulfite at a time. It exhibited the expected saturation kinetics and was catalyzed by the mitochondrial sulfite oxidase which we partially purified and recognized as a hemoprotein.[5] We kept returning to this real sulfite oxidase, over the years, and ultimately purified it to homogeneity and discovered that it contained molybdenum as well as heme.[6-8] It is an interesting and metabolically important enzyme. Humans, so unfortunate as to be born with a defect in it, suffer grave and ultimately fatal disabilities.[9,10] Nevertheless, the real sulfite oxidase was not on one of the trails to SOD. It was, in this sense, merely a diversion and we shall say no more about it.

The second kind of sulfite oxidation, catalyzed by liver extracts, was a free radical chain reaction, which did ultimately lead us to SOD. This second kind of sulfite oxidation went faster the greater the concentration of sulfite, with no indication of substrate saturation, because the length of individual reaction chains could increase, almost without limit, with increasing concentration of sulfite. When the liver extracts were dialyzed, they lost the ability to catalyze this free radical chain oxidation of sulfite. There was obviously a low-molecular-weight cofactor and I recall hoping that someday I would know what it was. Its isolation from several kilos of beef liver was laborious, but straightforward and soon I knew that it was hypoxanthine.

Hypoxanthine directly suggested the enzyme which oxidizes it, namely xanthine oxidase; xanthine oxidase, when acting on hypoxanthine, was found to initiate the free radical chain oxidation of sulfite.[11] One free radical often begets another and it seemed likely that xanthine oxidase was generating a free radical which, in turn, was initiating the chain oxidation of sulfite. The nature of this enzyme-generated free radical was not immediately obvious and if we could not understand, we could at least use the phenomena we had found. Consequently we used the chain oxidation of sulfite as a sensitive means for detecting the production of reactive radicals by illuminated dyes[12] and by several enzymes.[13]

Xanthine oxidase reduces molecular oxygen to hydrogen peroxide and, since hydrogen peroxide did not initiate sulfite oxidation, it seemed possible that there was an unstable precursor of H_2O_2 which could do so. The precursor of the two-electron reduced form of molecular oxygen should be the one-electron reduced form and that is the superoxide radical ($O_2^{\cdot-}$) well known to radiation chemists, but at the time of no interest to biochemists. Could $O_2^{\cdot-}$ initiate sulfite oxidation? A pair of platinum electrodes were sealed through the glass walls of a Warburg flask and the oxidation of sulfite was followed, by microrespirometry, as a function of an imposed DC potential. The electric current did cause sulfite oxidation, and the rate of sulfite oxidation, plotted against imposed potential, duplicated the known polarographic reduction wave for molecular oxygen. $O_2^{\cdot-}$ could thus initiate sulfite oxidation and we supposed that it was the product of the aerobic xanthine oxidase reaction which was reponsible for the initiation of the free radical chain oxidation of sulfite.[13] The first trail had led us to superoxide production by an oxidative enzyme.

IV. THE CYTOCHROME *c* TRAIL

This second long journey actually began with a paper by Horecker and Heppel.[14] They reported that xanthine oxidase could reduce cytochrome *c* but only in the presence of oxygen. Molecular oxygen and cytochrome *c* could ordinarily be thought of as competing electron acceptors, so an oxygen requirements for cytochrome *c* reduction is very peculiar. They tentatively suggested that xanthine oxidase reduced O_2 to H_2O_2, which was then reducing the cytochrome *c*. We tested this proposal in several ways and it was clear that H_2O_2 was not responsible for the reduction of cytochrome *c* by the aerobic xanthine oxidase reaction.[15] If H_2O_2 was not itself the reductant of cytochrome *c*, wasn't it likely that the unstable one-electron reduced form of O_2 was the reductant? Once again a research trail had led to the production of $O_2^{\cdot-}$ by xanthine oxidase. But how did SOD enter the story?

V. CYTOCHROME *c* AND THE SOD CONNECTION

There was some disagreement in the literature concerning the ability of xanthine oxidase to reduce cytochrome *c*. Some workers found that cytochrome *c* was readily reduced,[14,16] while others found cytochrome *c* to be a very poor electron acceptor, in comparison to molecular oxygen.[17,18] Our approach to this problem was to test several different preparations of cytochrome *c*. Some were rapidly reduced, whereas others were not. Moreover, the preparations which were not readily reduced, by xanthine oxidase, could inhibit the reduction of those that were. It was apparent that cytochrome *c* was sometimes contaminated with a potent, heat-labile, inhibitor of the reduction of cytochrome *c* by xanthine oxidase. This inhibitor was strictly competitive with cytochrome *c* and was then incorrectly identified as myoglobin.[19]

Given the 20/20 vision of hindsight, it is now perfectly clear that the thermolabile inhibitor of cytochrome *c* reduction was SOD, which could compete with cytochrome *c* for the flux of $O_2^{\cdot-}$ generated by xanthine oxidase. Given what we now know about the catalytic potency of SOD, it is also clear that our most contaminated preparation of cytochrome *c* had less than 1% SOD in it. This information was hard to come by. What was then clear was that O_2 was somehow mediating the transfer of electrons from xanthine oxidase to cytochrome *c* and that $O_2^{\cdot-}$ was the probable intermediate form of O_2. Verbal proposal of the enzymatic production of superoxide radical was met by extreme skepticism, if not derision, so in our written report, we proposed that $O_2^{\cdot-}$ remained bound to xanthine oxidase.[15] This provided a satisfyingly consistent, if incorrect, explanation. Thus, xanthine oxidase could reduce O_2 to $O_2^{\cdot-}$ which, while bound to the enzyme, could initiate sulfite oxidation or reduce cytochrome *c*. The thermolabile, competitive inhibitor of cytochrome *c* reduction could bind over the $O_2^{\cdot-}$ generating site, in competition with cytochrome *c*. It could thus prevent the reduction of cytochrome *c* or the initiation of sulfite oxidation without interfering with the catalytic turnover of xanthine oxidase because, in its presence, the bound $O_2^{\cdot-}$ could be further reduced to H_2O_2 and released from the enzyme. It was McCord who finally and decisively disposed of the notion that $O_2^{\cdot-}$ remained bound to xanthine oxidase and who thus proved that the thermolabile inhibitor of cytochrome *c* reduction and of sulfite oxidation was a superoxide dismutase. But that quantum leap in understanding did not occur until 1968, by which time we had walked many miles along the third trail to SOD.

VI. TRAIL THREE — THE CHEMILUMINESCENCE BYWAY

Totter et al.[20,21] had noted that the aerobic xanthine oxidase reaction could elicit chemiluminescence from compounds such as lucigenin and had proposed that oxygen radicals were involved.[22] This seemed like another useful way to explore the properties of the enzyme-bound superoxide radical. Compounds which had already been seen to block the $O_2^{\cdot-}$-mediated reduction of cytochrome *c* and the initiation of sulfite oxidation were also effective in blocking the induction of chemiluminescence from lucigenin or from luminol. We concluded that $O_2^{\cdot-}$, generated by xanthine oxidase, was responsible for inducing these chemiluminescences.[23] Moreover, since rabbit liver aldehyde oxidase and a bacterial dehydroorotic dehydrogenase, with their substrates, could replace the xanthine oxidase reaction, it was apparent that $O_2^{\cdot-}$ production was not a peculiarity of xanthine oxidase but might be a widespread phenomenon. The view from the luminescence trail was thus consistent with that already seen from the sulfite and from the cytochrome *c* trails. This increased our self confidence, and self confidence was badly needed by anyone then venturing to propose that enzymes might be producing superoxide radicals.

VII. A SABBATICAL AND A FIVE-YEAR DIVERSION

Had we been wise enough, we could have gone on to discover SOD in 1963. All of the background information was in hand and it was trying to tell us something important, but we were not listening hard enough. It is probably a rationalization to blame this on the noise generated by extraneous events. There was a sabbatical year spent with Westheimer at Harvard and it did ignite an interest in another enzyme, i.e., acetoacetic decarboxylase. There were fascinating anion effects on the activity of this enzyme[24] and the opportunity did arise to specifically label its active site.[25] One infidelity to xanthine oxidase and superoxide led to another and soon we were thoroughly sidetracked. We dallied with such lovely problems as the stimulation of horseradish peroxidase by nitrogenous ligands,[26] the competitive inhibition of xanthine oxidase by urea and guanidinium ion,[27] the reaction of boric acid with β-keto acids,[28] and others, many others. We went so far astray as to contemplate the competitive inhibition of uricase by *s*-triazines.[29] Indeed, we might never have returned to the superoxide fold, were it not for a paper by Pocker and Meany[30] dealing with carbonic anhydrase.

VIII. SAVED BY CARBONIC ANHYDRASE

Pocker and Meany discovered that carbonic anhydrase was not specific for CO_2, but would catalyze the hydration of other carbonyl compounds, such as acetaldehyde.[30] In the course of their studies they noted that the rate of hydration of acetaldehyde was conveniently low at low ionic strength, was generally[14] acid-base catalyzed, and exhibited an equilibrium constant of approximately 2.3 at 0°C. The textbook version of the enzymic oxidation of aldehydes at that time incorporated hydration to the gem diol, which was then oxidized to the acid. The data published by Pocker and Meany suggested a means for testing this mechanism.

In essence, the anhydrous aldehyde substrate would be rapidly added to the enzymic reaction mixture, under conditions where rate was a function of substrate concentration. If the gem diol hydrate was the true substrate, then the rate would initially be slow, but would accelerate as the hydration reaction proceeded towards equilibrium. On the contrary, if the free aldehyde was the true substrate, the rate would progressively decelerate as the aldehyde was converted to the hydrate. An enzyme with a high K_m for acetaldehyde was preferred and a means of spectrophotometrically monitoring the rate of oxidation was considered essential. Xanthine oxidase was chosen as the test enzyme and the reduction of cytochrome *c* was selected as the spectrophotometric signal to be followed.

These experiments were actually done and the results left no doubt but that the free aldehyde, rather than the gem diol hydrate, was the substrate for xanthine oxidase.[31] It was gratifying to correct the textbook version of this reaction and we did take the trouble to generalize our conclusion by testing other aldehyde oxidizing enzymes[32] but the real importance of this project was that it put us firmly back onto the superoxide trail. Thus the carbonic anhydrase we had prepared to catalyze the hydration of acetaldehyde was contaminated with SOD and, as a consequence, it potently and competitively inhibited the reduction of cytochrome *c*,[33] much as myoglobin has previously been seen to do.[19] Furthermore, the inhibitory constant for the inhibition by carbonic anhydrase was a very impressive 2×10^{-9} *M* and a number of other proteins, chosen at random, were without effect. This simply could not be ignored.

IX. McCORD REMOVES THE ROADBLOCK

At this time we were still thinking of $O_2^{\overline{\cdot}}$ as an intermediate which remained bound to the xanthine oxidase. This led to the view that cytochrome *c* had to bind to the enzyme in order to interact with the bound $O_2^{\overline{\cdot}}$ and that competitive inhibitors of cytochrome *c* reduction were likewise binding to the xanthine oxidase. This misconception constituted a barrier which prevented us from understanding the true situation. This barrier was removed by Joe M. McCord. McCord had joined our group as a graduate student and was assigned the task of demonstrating the binding of carbonic anhydrase to xanthine oxidase and of directly measuring the binding constant. These were technically demanding manipulations and as they yielded consistently negative results, he grew understandably impatient with them.

On April 2, 1968 he was measuring the reduction of cytochrome *c* by the xanthine oxidase reaction. When the concentration of cytochrome *c* was varied, a classical hyperbolic saturation curve was generated. This is usually interpreted as an indication of the saturation of a binding site and is ordinarily converted into a dissociation constant of the presumed enzyme-substrate complex. McCord added methylene blue to mediate the reduction of cytochrome *c* by the enzyme. This dye was known to carry electrons from the enzyme to cytochrome *c*. Thus, in the absence of oxygen, native xanthine oxidase will not reduce cytochrome *c* and adding methylene blue will enable it to do so. Furthermore, there was no established belief that methylene blue remained bound to the enzyme. In the presence of methylene blue McCord again observed a classical saturation curve for cytochrome *c*, but this time the dissociation constant for the supposed xanthine oxidase-cytochrome *c* complex was different.

If one could change the "dissociation constant" for cytochrome *c* merely by adding an electron-carrying dye, then why believe that saturation kinetics implied binding? Furthermore, if methylene blue could shuttle between the enzyme and cytochrome *c*-carrying electrons, then why couldn't oxygen do likewise? The apparent saturation of the rate at high levels of cytochrome *c* was readily envisaged, in the absence of binding to the enzyme. Thus the reduced dye could either reduce cytochrome *c* or it could autoxidize. At low concentrations of cytochrome *c*, autoxidation would predominate and the rate of cytochrome *c* reduction would be low. As cytochrome *c* concentration was raised, it would capture an ever greater fraction of the electrons being carried by the dye. The rate would reach an upper limit or saturation, when the cytochrome *c* was intercepting every molecule of the reduced dye. Furthermore, the concentration of cytochrome *c* required to saturate the reaction, and therefore the K_s for cytochrome *c*, would depend upon the concentration of dye used. Exactly similar reasoning would apply when $O_2^{\overline{\cdot}}$, rather than reduced methylene blue, was the electron carrier. In the latter case, the reaction competing with the cytochrome *c* would not be an autoxidation, but rather a dismutation, i.e., $O_2^{\overline{\cdot}} + O_2^{\overline{\cdot}} + 2H^+ \rightarrow H_2O_2 + O_2$.

On April 3, 1968, without revealing what he had in mind, McCord asked me if it was true that in the case of binding of a substrate to an enzyme the K_m for that substrate was independent of the concentration of the enzyme. I responded in the affirmative, noting the necessary condition that the concentration of the substrate must greatly exceed that of the enzyme. McCord was satisfied with that answer and proceeded to measure K_s for cytochrome *c* as a function of the concentration of xanthine oxidase. This K_s for cytochrome *c* was not independent of the concentration. Quite the contrary, it was proportional to the square root of the enzyme concentration. The K_s for the substrate xanthine, which does bind to the enzyme, was, as expected, independent of the concentration of xanthine oxidase.

The roadblock had been suddenly and completely removed and our blinders were off. We could, in that instant, understand the true situation. Cytochrome *c* did not bind to xanthine oxidase. It was reduced in free solution by $O_2^{\overline{\cdot}}$ released from that enzyme. Furthermore, the proteins which had been seen to competitively inhibit the reduction of cytochrome *c* must operate by intercepting $O_2^{\overline{\cdot}}$ in free solution. They were active in catalytic amounts, so they must have been operating to facilitate the dismutation of $O_2^{\cdot -}$. The concept of superoxide dismutase was thus born. If all of this were correct, then the SOD activity of carbonic anhydrase or of myoglobin should also be able to prevent the initiation of sulfite oxidation by $O_2^{\cdot -}$ generated at a platinum electrode. The old Warburg vessel with the sealed-in electrodes[13] was disinterred and put into use in affirming this prediction. By the time the paper describing all of this was in galley proof, the SOD activity had been partially purified and recognized as being distinct from carbonic anhydrase. This was added as a note to the paper which, more than any other, opened the way to our present understanding of SOD.[34]

X. THE FIRST ISOLATION OF SOD

The inhibition of the reduction of cytochrome *c* during the xanthine oxidase reaction proved to be a convenient assay for SOD. One unit of SOD activity was arbitrarily defined as that amount which caused 50% inhibition under specified conditions. With this assay as his guide, McCord isolated the SOD of bovine erythrocytes and obtained a blue-green copper-containing enzyme, which was recognized as being identical to the hemocuprein and erythrocuprein, which had already been isolated and studied by others, but without recognition of its enzymic activity. In order to be absolutely certain of the action of SOD on $O_2^{\cdot -}$, the xanthine oxidase system was replaced by electrochemically synthesized tetrabutyl ammonium superoxide and cytochrome *c* was replaced by tetranitromethane and by epinephrine. In all cases, SOD catalytically scavenged $O_2^{\cdot -}$ and thus interfered with the $O_2^{\cdot -}$-dependent process being measured. The Cu was shown to be essential for this enzyme to function, because its reversible removal was associated with a reversible loss of activity. The cytochrome *c* reduction assay was applied to several tissues and SOD activity was evident in all of them. All of this was reported in the paper describing the first isolation of SOD.[35]

XI. AN EXPLOSION OF LITERATURE

A good workman can hardly resist the temptation to use first-rate tools and we had such tools in our hands. Thus, the cytochrome *c* reduction assay made it easy to look for SOD in diverse places, and the availability of a pure SOD made it easy to detect $O_2^{\cdot -}$. We followed our inclinations and used these tools. $O_2^{\cdot -}$ production was found to be a commonplace event[36] and the amounts produced by xanthine oxidase were seen to represent a sizable fraction of the total oxygen reduction.[37] $O_2^{\cdot -}$ plus H_2O_2 was seen to generate $OH\cdot$,[38] a process now known to be dependent upon iron present in the phosphate buffers used and quite possibly of physiological significance. We isolated SOD from *E. coli* which contained manganese,[39] and we devised a new assay which could be applied to polyacrylamide gels as an activity stain for SOD.[40] We did many things and for a while we had the whole marvelous playground to ourselves. There were, alas, alert investigators in many fields who recognized the utility of these tools to their own projects and who soon began to publish exciting findings. There is an autocatalysis built into the scientific enterprise, as each new report stimulates additional experimental work. As a result the superoxide literature has burgeoned to the point where a complete bibliography would now include over 6000 references. What started out as a peaceful, if lonely, trail-walk has developed into a jostling big city scramble. The magnitude of this growth during the past decade is partially reflected by the chapters which follow. I hope you will find them interesting and instructive.

ACKNOWLEDGMENT

I had the good fortune, during the early stages of this work, to have as my mentor, Dr. Philip Handler, who although beset by many other cares, was always willing to listen to and to carefully consider the import of the day-to-day "bench" results. His incredible memory banks and wise counsel helped remove many rocks from the trails.

REFERENCES

1. **Handler, P., Bernheim, M. L. C., and Klein, J. R.,** The oxidative demethylation of sarcosine to glycine, *J. Biol. Chem.*, 138, 211, 1941.
2. **Heimberg, M., Fridovich, I., and Handler, P.,** The enzymatic oxidation of sulfite, *J. Biol. Chem.*, 204, 913, 1953.
3. **Fridovich, I. and Handler, P.,** Further studies of sulfite oxidase, *Fed. Proc.*, 13, 212, 1954.
4. **Backstrom, H. L. J.,** The chain-reaction theory of negative catalysis, *J. Am. Chem. Soc.*, 49, 1460, 1927.
5. **MacLeod, R. M., Farkas, W., Fridovich, I., and Handler, P.,** Purification and properties of hepatic sulfite oxidase, *J. Biol. Chem.*, 236, 1841, 1961.
6. **Cohen, H. J. and Fridovich, I.,** Hepatic sulfite oxidase, purification and properties, *J. Biol. Chem.*, 246, 359, 1971.
7. **Cohen, H. J. and Fridovich, I.,** Hepatic sulfite oxidase, the nature and function of the heme prosthetic groups, *J. Biol. Chem.*, 246, 367, 1971.
8. **Cohen, H. J., Fridovich, I., and Rajagopalan, K. V.,** Hepatic sulfite oxidase, a functional role for molybdenum, *J. Biol. Chem.*, 246, 374, 1971.
9. **Johnson, J. L. and Rajagopalan, K. V.,** Purification and properties of sulfite oxidase from human liver, *J. Clin. Invest.*, 58, 543, 1976.
10. **Johnson, J. L. and Rajagopalan, K. V.,** Human sulfite oxidase deficiency. Characterization of the molecular defect in a multicomponent system, *J. Clin. Invest.*, 58, 551, 1976.
11. **Fridovich, I. and Handler, P.,** Xanthine oxidase. III. Sulfite oxidation as an ultrasensitive assay, *J. Biol. Chem.*, 233, 1578, 1958.
12. **Fridovich, I. and Handler, P.,** Detection of free radicals in illuminated dye solutions by the initiation of sulfite oxidation, *J. Biol. Chem.*, 235, 1835, 1960.
13. **Fridovich, I. and Handler, P.,** Detection of free radicals generated during enzymic oxidations by the initiation of sulfite oxidation, *J. Biol. Chem.*, 236, 1836, 1961.
14. **Horecker, B. L. and Heppel, L. A.,** The reduction of cytochrome *c* by xanthine oxidase, *J. Biol. Chem.*, 178, 683, 1949.
15. **Fridovich, I. and Handler, P.,** Xanthine oxidase. V. Differential iniibition of the reduction of various electron acceptors, *J. Biol. Chem.*, 237, 916, 1962.
16. **Weber, M. M., Lenhoff, H. M., and Kaplan, N. O.,** The reduction of inorganic iron and cytochrome *c* by flavin enzymes, *J. Biol. Chem.*, 220, 93, 1956.
17. **Westerfeld, W. W., Richert, D. A., and Higgins, E. S.,** Further studies with xanthine oxidase inhibitors, *J. Biol. Chem.*, 234, 1897, 1959.
18. **Avis, P. G., Bergel, F., and Bray, R. C.,** Estimation of the cofactors and the catalytic activities of enzyme fractions from cow milk, *J. Chem. Soc.*, 1219, 1956.
19. **Fridovich, I.,** Competitive inhibition by myoglobin of the reduction of cytochrome *c* by xanthine oxidase, *J. Biol. Chem.*, 237, 584, 1962.
20. **Totter, J. R., Scoseria, J. L., and Medina, V. J.,** Systems artificiales de luciferina-luciferase, *An. Fac. Med. Montevideo*, 44, 463, 1959.
21. **Totter, J. R., Medina, V. J., and Scoseria, J. L.,** Luminescence during the oxidation of hypoxanthine by xanthine oxidase in the presence of dimethylbisacridylium nitrate, *J. Biol. Chem.*, 235, 238, 1960.
22. **Totter, J. R., deDugros, E. C., and Riveiro, C.,** The use of chemiluminescent compounds as possible indicators of radical production during xanthine oxidase action, *J. Biol. Chem.*, 235, 1839, 1960.
23. **Greenlee, L., Fridovich, I., and Handler, P.,** Chemiluminescence induced by the operation of iron-flavoproteins, *Biochemistry*, 1, 779, 1962.
24. **Fridovich, I.,** Inhibition of acetoacetic decarboxylase by anions — the Hofmeister lyotrophic series, *J. Biol. Chem.*, 238, 592, 1963.

25. **Fridovich, I. and Westheimer, F. H.**, Studies on the mechanism of acetoacetic decarboxylase, *J. Am. Chem. Soc.*, 84, 3208, 1962.
26. **Fridovich, I.**, The stimulation of horseradish peroxidase by nitrogenous ligands, *J. Biol. Chem.*, 238, 3921, 1963.
27. **Fridovich, I.**, Competitive inhibition of xanthine oxidase by urea and guanidinium ion, *J. Biol. Chem.*, 239, 3519, 1964.
28. **Neece, S. and Fridovich, I.**, Boric acid reaction with β-keto acids and its use in assaying acetoacetic decarboxylase, *Arch. Biochem. Biophys.*, 108, 240, 1964.
29. **Fridovich, I.**, The competitive inhibition of uricase by oxonate and by related derivatives of s-triazines, *J. Biol. Chem.*, 240, 2491, 1965.
30. **Pocker, Y. and Meany, J. E.**, The catalytic versatility of carbonic anhydrase from erythrocytes. The enzyme-catalyzed hydration of acetaldehyde, *J. Am. Chem. Soc.*, 87, 1809, 1965.
31. **Fridovich, I.**, The mechanism of the enzymatic oxidation of aldehydes, *J. Biol. Chem.*, 241, 3126, 1966.
32. **Naylor, J. F., III and Fridovich, I.**, The form of the aldehyde susceptible to enzymic oxidation, *J. Biol. Chem.*, 243, 341, 1968.
33. **Fridovich, I.**, A reversible association of bovine carbonic anhydrase with milk xanthine oxidase, *J. Biol. Chem.*, 242, 1445, 1967.
34. **McCord, J. M. and Fridovich, I.**, The reduction of cytochrome *c* by milk xanthine oxidase, *J. Biol. Chem.*, 243, 5753, 1968.
35. **McCord, J. M. and Fridovich, I.**, Superoxide dismutase — an enzymic function for erythrocuprein (hemocuprein), *Biol. Chem.*, 244, 6049, 1969.
36. **McCord J. M. and Fridovich, I.**, The utility of superoxide dismutase in studying free radical reactions. II. The mechanism of the mediation of cytochrome *c* reduction by a variety of electron carriers, *J. Biol. Chem.*, 245, 1374, 1970.
37. **Fridovich, I.**, Quantitative aspects of the production of superoxide anion radical by milk xanthine oxidase, *J. Biol. Chem.*, 245, 4053, 1970.
38. **Beauchamp, C. and Fridovich, I.**, A mechanism for the production of ethylene from methional: the generation of hydroxyl radical by xanthine oxidase, *J. Biol. Chem.*, 245, 4641, 1970.
39. **Keele, B. B., Jr., McCord, J. M., and Fridovich, I.**, Superoxide dismutase from *Escherichia coli* B; a new manganese-containing enzyme, *J. Biol. Chem.*, 245, 6176, 1970.
40. **Beauchamp, C. and Fridovich, I.**, Superoxide dismutase: improved assays and an assay applicable to polyacrylamide gels, *Anal. Biochem.*, 44, 276, 1971.

Chapter 2

SUPEROXIDE DISMUTASES: PROTEIN CHEMISTRY AND STRUCTURE-FUNCTION RELATIONSHIPS

Howard M. Steinman

TABLE OF CONTENTS

I. INTRODUCTION

In 1969, the biochemical community was introduced to a new enzyme, conceived in mystery and dedicated to the abolition of superoxide free radicals.[1] Novel in its free radical substrate, facile in its purification, and remarkable in its stability, superoxide dismutase (SOD) was seized upon by biochemists of enzymological, physical, and biological bents. The vigor which this discovery instilled into biological studies of oxygen-derived free radicals is manifest in subsequent chapters of this monograph. However, the alacrity with which physiological and medical bailiwicks have been penetrated belies the fact that the phenomenon of SOD began in a quite conventional, enzymological gait. To wit, purification of the first dismutase from bovine erythrocytes was duly followed by purifications and characterizations of enzymes from the customary zoological repertoire, which led in turn to descriptions of the three metalloenzyme forms: dismutases containing both copper and zinc, and dismutases containing either manganese or iron as co-factors.

The objective of this chapter is to elaborate upon the structural and functional properties of these three forms, deduced from studies in which SODs have been construed less as protagonists against oxygen toxicity or potential therapeutic agents, and more as enzymes, whose structural and mechanistic properties are of interest in their own right. The presentation is comparative, juxtaposing enzymes of different metal form, and comparing enzymes of the same form, isolated from different species.

The phylogenetic diversity of SODs is rather beguiling. The three forms segregate into two families, copper-zinc and manganese/iron dismutases, of dramatically different properties. These two families apparently evolved from different ancestral proteins and arose in different eras during the origin of species.[2] What selection pressures necessitated the evolution of the copper-zinc family, when a manganese/iron family, of comparable catalytic efficiency, was already in existence? What are the physiological features of present-day organisms which underlie the different distribution of these metalloenzyme forms among species, and between cytosolic and organelle locales within a single cell? One goal of this chapter is to begin to assemble information about the comparative structural and functional properties of dismutases from which such questions may ultimately be answered.

In this chapter, the concern for encompassing dismutases from a variety of species has been tempered by an effort to explain the rationale and strategy behind the truly varied experimental approaches that have been recruited in studies of dismutase structure and function. Several reviews and monographs dealing with structure-function relationships among SODs have appeared in recent years.[3-8] (The literature search for this review was completed in December 1980.)

II. A SURVEY OF SUPEROXIDE DISMUTASES

A. Introduction

A description of the general properties of the three metalloprotein forms is necessary, before delving into the detailed structure-function studies which have been made upon specific enzymes. The objective of this section is to define the distinguishing gross macromolecular traits of the three forms, to summarize their distribution among prokaryotic and eukaryotic species, and to discuss briefly some of the biological queries which have been stimulated by phylogenetic studies of SODs. A compilation of some of these properties for purified dismutases of the Cu-Zn and the Mn/Fe families are listed, respectively, in Tables 1[1,9-36] and 2.[9,11,37-74]

B. Molecular Weight, Subunit Structure, and Metal Content

1. Copper-Zinc Dismutases

The copper-zinc dismutases are a remarkably conserved family, with respect to their gross structural properties.[4,75,76] Without exception, the purified enzymes have been shown to be dimers (native molecular weight usually 31,000 to 33,000) containing very nearly 2.0 g-atoms of copper and 2.0 g-atoms of zinc per mole. The two subunits are apparently identical and are associated solely by noncovalent interactions. Subunit molecular weights estimated by physical methods (15,000 to 17,000) are in good agreement with the results of the complete amino acid sequences, now known for the enzymes from bovine[77-79] and human erythrocytes,[80,81] from yeast,[82,83] and from horse liver[84] (subunit molecular weights 15,600 to 16,000). These sequence data and the crystal structure of the bovine erythrocyte enzyme[85-87] demonstrate that each subunit contains one intrachain disulfide bond and one active site, containing an atom of copper and an atom of zinc. (Details of these structural characterizations and their functional significance are given in Section III.)

The literature does contain reports of exceptions to these common properties, but in all cases, subsequent publications have suggested that the initial report may have been erroneous. Nonidentity of the subunits was originally reported for the dismutase from human erythrocytes[88] and *Photobacterium leiognathi*,[10] however more recent work has suggested that this human[80,81] and bacterial dismutase[11] is composed of identical polypeptide chains. It was initially thought that the two subunits of the bovine erythrocyte enzyme were joined by disulfide bond(s).[24] In sodium dodecyl sulfate-polyacrylamide gel electrophoresis, the apparent molecular weight of the holoenzyme

Table 1
COPPER-AND-ZINC-CONTAINING SODs — SUMMARY OF STRUCTURAL PROPERTIES

Species	Native mol wt	Subunit structure	Metals (g-atoms/ subunit)	Comments	Ref.
Prokaryotes					
Paracoccus denitrificans	ND	ND	Cu +, Zn +	Purified without $CHCl_3$-EtOH	9
Photobacterium leiognathi	33,100	Dimer	Cu 0.5, Zn 1.3	Nonidentical subunits	10
Photobacterium leiognathi	32,000	Dimer	ND	Identical subunits	11
Fungi					
Fusarium oxysporum	ND	ND	ND	Purified without $CHCl_3$-EtOH	12
Neurospora crassa	32,200	ND	Cu 1.0, Zn 0.9	Purified without $CHCl_3$-EtOH	12
Neurospora crassa	31,100	Dimer	Cu 1.0, Zn 0.9		13
Saccharomyces cerevisiae	32,700	Dimer	Cu 1.0, Zn 0.9		14
Saccharomyces cerevisiae	30,500	Dimer	Cu 0.9, Zn 1.0		15,16
Seed plants					
Green pea	31,500	ND	Cu 1.0, Zn 1.0	Purified without $CHCl_3$-EtOH	17
Spinach leaf	32,200	Dimer	Cu 1.1, Zn 1.1	Purified without $CHCl_3$-EtOH	18
Wheat germ isozyme I	31,000	Dimer	Cu 0.9, Zn 1.0	Isozymes differ in stability to SDS	19
Wheat germ isozyme II	31,000	Dimer	Cu 0.8, Zn 1.0	SDS	19
Fish					
Atlantic croaker	31,000	Dimer	ND		11
Black sea bass	31,000	Dimer	ND		11
Pony fish	32,000	Dimer	ND		11
Red snapper	32,000	Dimer	ND		11
Speckled trout	32,000	Dimer	ND		11
Swordfish liver	32,500	Dimer	Cu 1.0, Zn 1.0		20
White marlin	31,000	Dimer	ND		11
Other vertebrates					
Boar spermatozoa	30,000	Dimer	Cu 0.9, Zn 1.0	Purified without $CHCl_3$-EtOH	21
Bovine brain	ND	ND	Cu 0.8	Purified without $CHCl_3$-EtOH	22
Bovine erythrocytes	32,600	ND	Cu 1.0	Discovery of dismutase	1
Bovine erythrocytes	33,000	ND	Cu 1.0, Zn 1.0		23
Bovine heart	32,500	Dimer	Cu 0.8, Zn 0.9		24
Bovine liver	32,000	ND	Cu 1.0, Zn 0.9		16
Bovine spermatozoa	33,000	Dimer	ND	Purified without $CHCl_3$-EtOH	25
Chicken liver	30,600	Dimer	Cu 0.9, Zn 0.8	Reactive - SH group	26
Horse liver	33,000	Dimer	Cu 1.0, Zn 1.0	Purified without $CHCl_3$-EtOH	27
Human brain	35,700	ND	Cu 1.0	Purified without $CHCl_3$-EtOH	28
Human erythrocytes	34,600	ND	Cu 1.0	Purified without $CHCl_3$-EtOH	29
Human erythrocytes	ND	ND	Cu 0.9, Zn 0.8	Purified without $CHCl_3$-EtOH	30,31
Human erythrocytes SOD I	ND	ND	Cu 1, Zn 1	320 nm chromophore present	32
Human erythrocytes SOD II	ND	ND	Cu 1, Zn 1	320 nm chromophore absent	32

Table 1 (continued)
COPPER-AND-ZINC-CONTAINING SODs — SUMMARY OF STRUCTURAL PROPERTIES

Species	Native mol wt	Subunit structure	Metals (g-atoms/ subunit)	Comments	Ref.
Human erythrocytes	ND	Dimer	Cu 1.0, Zn 1.0	Genetic variant	33
Human liver	35,600	ND	Cu 0.9	Purified without $CHCl_3$-EtOH	28
Porcine erythrocytes	ND	Dimer	Cu +		34
Porcine liver	ND	Dimer	Cu +		34
Ram erythrocytes	34,000	Dimer	Cu 1, Zn 1		35
Ram spermatozoa	34,000	Dimer	ND	Purified without $CHCl_3$-EtOH	35
Rat liver	31,000	ND	Cu 1.0, Zn 1.0		36

Note: ND, not determined; +, present, but not quantitated.

was approximately equal to the native molecular weight, if the protein was not treated with 2-mercaptoethanol. Following treatment with mercaptoethanol, a polypeptide of subunit molecular weight was observed. This effect of mercaptoethanol is now attributed to reduction of intrachain disulfide bonds.[19,76,79] The reduction destabilizes the structure of the individual subunit, which weakens subunit interactions and facilitates the dissociation of the holoenzyme. In contrast, the apoprotein is readily dissociated by sodium dodecyl sulfate, without treatment with mercaptoethanol. The conclusion drawn from these observations is that both the intrasubunit disulfide bonds and the metal-binding site(s) contribute to the strength of the subunit association.[76,79] The original misconception about the action of mercaptoethanol is but one example of the perplexities which have arisen from the unusual stability of the holoenzyme form of these enzymes (for another example, see Section III. F. 4).

The copper-zinc dismutases initially seemed to be a family of acidic proteins, because proteins from the mammalian species, first surveyed, had isoelectric points between 4 and 6.[21,23,27,29] It is now evident that this is not a pervasive trait, because several Cu-Zn enzymes from teleost fish are quite basic with pI values as high as 8.7.[11] While differing in their content of charged amino acid residues, Cu-Zn dismutases show two rather constant and characteristic features in their amino acid compositions. First, the content of tyrosine and tryptophan is very low; among the four enzymes whose sequences are known, one has a single tryptophan,[80,81] two have a single tyrosine,[78,82,83] and one has neither amino acid,[84] per polypeptide chain of about 150 to 155 residues. Amino acid compositions reported from analysis of other members of this family are consistent with these data, and the UV absorption spectra reflect the paucity of the strongly absorbing aromatic residues. Second, the content of glycine is high, from about 13 to 17 mol%, corresponding to approximately 20 to 26 residues per subunit chain, or an average of about one glycine per every 6 to 8 amino acid residues.[78,80-84] Glycine is distributed rather evenly throughout the chain in the four known sequences. In the three-dimensional structure of the bovine enzyme (Section III. B. 1)[85,86] this high content of glycine appears to be essential for the extensive β-pleated sheet conformation, with intervening sharp bends in the polypeptide, and it is probably for this reason that this generous endowment with glycine has been conserved throughout the evolutionary history of Cu-Zn dismutases.

2. Manganese/Iron Dismutases

Enzymes of the manganese/iron family are not as uniform in such basic properties

Table 2
MANGANESE- AND IRON-CONTAINING SODs — SUMMARY OF STRUCTURAL PROPERTIES

Species	Native mol wt	Subunit structure	Metals (g-atoms/ subunit)	Ref.
Prokaryotes: obligate anaerobes				
Chlorobium thiosulphatophilum	43,000	Dimer	Fe 0.90	37
Chromatium vinosum	41,000	Dimer	Fe 0.95	38
Desulfovibrio desulfuricans	43,000	Dimer	Fe 0.80	39
Prokaryotes: facultative and obligate aerobes, photosynthetic algae				
Bacillus megaterium	40,300	Dimer	Fe 0.5	41
Bacillus megaterium	ND	ND	Mn +	11,42
Bacillus stearothermophilus	40,000	Dimer	Mn 0.45	43,44
Escherichia coli B	39,500	Dimer	Mn 0.85	45
Escherichia coli B	ND	ND	Mn 0.60	46
Escherichia coli B	ND	ND	Mn 0.35—0.90	47
Escherichia coli B	38,700	Dimer	Fe 0.50	48
Escherichia coli B	ND	ND	Fe 0.90	49
Escherichia coli B hybrid	37,000	ND	Fe 0.40	50
Mycobacterium lepraemurium	45,000	Dimer	Mn 0.65	51
Mycobacterium phlei	80,000	Tetramer	Mn 0.87	52
Mycobacterium (Takeo)	61,500	Trimer	Mn 0.57	53
Mycobacterium tuberculosis	88,000	Tetramer	Fe 0.93	54
Paracoccus denitrificans	41,000	Dimer	Fe 1.0	9
Photobacterium leiognathi	41,000	Dimer	Fe 0.80	11, 55
Photobacterium sepia	41,000	Dimer	Fe 0.72	55
Plectonema boryanum	41,700	Dimer	Fe 0.98	56
Plectonema boryanum	36,500	Dimer	Fe 0.47	57
Pseudomonas ovalis	40,800	Dimer	Fe 0.55	58
Rhodopseudomonas spheroides	37,400	Dimer	Mn 0.55	40
Spirulina platensis	37,400	Dimer	Fe 0.49	40,59
Streptococcus faecalis	45,300	Dimer	Mn 0.65	60
Streptococcus mutans isozyme I	40,300	Dimer	Mn 0.93	61
Streptococcus mutans isozyme II	ND	Dimer	ND	61
Thermus aquaticus	80,000	Tetramer	Mn 0.53	62
Thermus thermophilus HB8	82,000	Tetramer	Mn 0.56	63
Eukaryotes				
Porphyridium cruentum (red alga)	40,000	Dimer	Mn 0.48	64
Euglena gracilis isozyme I	45,000	Dimer	Fe 0.90	65
Euglena gracilis isozyme II	45,000	Dimer	Fe 0.95	65
Pleurotus olearius c form	75,500	Tetramer	Mn 0.53	66
Pleurotus olearius m form	78,000	Tetramer	Mn 0.48	67
Saccharomyces cerevisiae	97,000	Tetramer	Mn 0.95	68
Brassica campestris (mustard leaves)	41,000	Dimer	Fe 0.80	69
Pisum sativum (pea)	84,600	Tetramer	Mn 0.25	70
Chicken liver	79,400	Tetramer	Mn 0.53	26
Rat liver	89,000	Tetramer	Mn 0.92	71
Bovine heart	90,000	Tetramer	Mn 0.47—0.76	72
Human liver	85,300	Tetramer	Mn 0.98	73,74

Note: ND, not determined; +, present, but not quantitated.

as molecular weight, subunit structure, and metal content.[4,76] Some are dimers; others are tetramers. The contents of the metal range from about 0.5 to 1 g-atom/mol of subunit polypeptide chain (see Table 2). Only one type of metal is present, manganese or iron. Regardless of subunit structure, the constituent polypeptide chain of these dismutases is larger than that of copper-zinc dismutases. From physical measurements, subunit molecular weights of 19,000 to 22,000 g/mol are obtained, in agreement with values of 22,000 and 23,000 g/mol for the polypeptide of the two enzymes of this family for which complete sequence data are available, the dimeric Mn-containing en zymes from *Bacillus stearothermophilus* [44] and *Escherichia coli* [89] Neither of these two enzymes contain half-cystine; the two subunits are identical in each dimer.

Although Mn/Fe dismutases are generally dimeric, other multimeric states exist. Tetramers have been identified in (1) all manganese-containing enzymes from eukaryotes, (2) manganese enzymes from three bacteria, *Mycobacterium phlei,*[52] *Thermus aquaticus,*[62] and *Thermus thermophilus,*[63] and (3) one iron-containing dismutase, that from *Mycobacterium tuberculosis.*[54] One trimer has been identified, in the manganese enzyme from *Mycobacterium* (Takeo).[53] The native molecular weight of the trimer, 62,000 to 65,000, appears to have been rigorously determined, by sedimentation equilibrium, with corrections for the dependence upon protein concentration.

The amino acid compositions of Mn/Fe enzymes are lower in glycine content and higher in the amounts of tyrosine and tryptophan than compositions of Cu-Zn dismutases. The amino acid compositions of the three different dismutase forms have been classified using rank order correlation and discriminant functions analysis.[11] The compositional properties of each individual metalloenzyme form are statistically distinct from one another. In fact, a dismutase can be classified as copper-zinc, manganese- or iron-containing on the basis of its amino acid composition alone.[11] Most of the Mn/Fe enzymes are acidic proteins, pI 4 to 5.[9,11,37,38,50,55,59,64,65,69,70] Higher isoelectric points have been reported for the manganese dismutases from *E. coli* B, pI = 6.6[50] and *Pleurotus olearius*, pI = 7.0.[66]

C. Purification and Stability Properties

1. Copper-Zinc Dismutases

a. Purification Properties

To the practicing biochemist, it is the stability differences which distinguish the two dismutase families more strikingly than the structural parameters cited above. These differences are perhaps confronted most clearly in the purification schemes by which most Cu-Zn enzymes have been purified. The crude homogenate is treated with a chloroform-ethanol mixture, and the supernate obtained after centrifugation is saturated with dibasic potassium phosphate, effecting a separation of organic and aqueous phases. The SOD partitions into the organic phase, from which it is precipitated by addition of acetone. The ability to withstand such harsh treatment classifies copper-zinc dismutases as one of the most stable enzyme families and also contributes enormously to the ease with which these enzymes are purified, because many other proteins are irreversibly denatured or precipitated by this treatment.[1,74]

The purification of many Cu-Zn dismutases has commenced with such a protocol: chloroform-ethanol treatment of the homogenate, followed by salting out of the organic phase, and acetone precipitation. The protocol, an adaptation of the Tsuchihashi method for removing hemoglobin from erythrocyte lysates,[90] will be referred to as the McCord-Fridovich method, after its use in isolation of the first SOD.[1]

Objections have been raised against the suitability of the McCord-Fridovich protocol in the purification of dismutase from human erythrocytes. To wit, the exposure to organic solvents has been impugned for the altered sedimentation coefficient and isoelectric point compared to the values for human dismutase, purified by a procedure which eschews organic solvents.[91,92] It should be noted that the procedure which was shown to irreversibly alter the S-value and pI differed from the McCord-Fridovich method. The phase separation with phosphate was omitted, and in lieu of the acetone precipitation, lead acetate was added directly to the chloroform-ethanol mixture to precipitate the enzyme.[93] Consequently, it is not clear whether the treatment with organic solvents alone was responsible for structural alterations, whether the culprit was actually the precipitation step with a heavy metal ion, or whether the combination of the two treatments.

The lead acetate precipitation has never been incorporated into the McCord-Fridovich protocol. In fact, human erythrocyte dismutase purified strictly by the protocol shows no alterations in electrophoretic character, metal content, or gross molecular shape, compared to human enzyme purified by a method which avoids any exposure to organic solvents.[30,31] It has been concluded that the McCord-Fridovich method is acceptable for purifying dismutase from human and other sources without gross structural changes or loss of activity.[30-32] It may be noted that not all Cu-Zn dismutase purifications have included the McCord-Fridovich protocol; some exceptions in nonhuman species appear in Table 1.[9,10,12,17,18,21,22,25,27,35]

The McCord-Fridovich method has been correlated with an alteration of a particularly reactive sulfhydryl group in human dismutase, now known to be Cys-111 in the amino acid sequence.[80,81] This sulfhydryl group has not been identified in a Cu-Zn dismutase from any other species, and its alteration does not influence enzyme activity. For some years, it has been known that the dismutase from human erythrocytes and other human tissues displays a small, but distinct, absorbance band in the vicinity of 320 to 325 nm.[32,32a] This band is absent from the absorption spectrum of other (eg., equine and bovine erythrocyte) Cu-Zn dismutases. The absorbance appears to arise from protein bound persulfide and protein$-CH_2-S-(S)_n-R$ ($n \geq 0$) groups, which are neither related to the catalytic activity nor near the active site. These unusual sulfur derivatives, most probably alternative forms of Cys-111, are artifacts of the McCord-Fridovich purification scheme. Both the absorbance band and the reactivity properties of persulfides are absent when human dismutase is purified by procedures which omit organic solvents.[30,31]

Among the four Cu-Zn dismutases whose amino acid sequences are known, the human enzyme is the only one with a cysteine at residue 111.[77-84] It is not known if Cys-111 is a recent evolutionary acquisition or if it is found in other primates. The Cu-Zn dismutase from chicken liver contains a single reactive (noncatalytic) sulfhydryl group per subunit, which has been implicated in the autoreduction and the polymerization of this enzyme.[26] It is not known if this residue is homologous to human Cys-111.

b. Other Stability Properties

The stability of the enzyme from bovine erythrocytes has been documented more extensively than that of any other Cu-Zn dismutase. The activity is stable in 8 *M* urea for hours at room temperature, in the presence or absence of ethylenediamine tetraacetate(EDTA).[94,95] It is neither inactivated nor dissociated by sodium dodecyl sulfate, and treatment for determination of subunit molecular weight by sodium dodecyl sulfate-polyacrylamide gel electrophoresis requires exposure to mercaptoethanol or EDTA, preferably with heating to 40 to 55°C., or higher, to insure dissociation.[79,94] A temperature of 67°C is required for inactivation of 50% of the activity in 10 min.[94] The activity (second-order rate constant for catalysis) is constant over a range from pH 4.5 to pH 9.5.[96] Furthermore, the bovine enzyme is stable to repeated freeze-thaw cycles, and to prolonged refrigeration.[74] The holoenzyme is inactivated by 6 *M* guanidine hydrochloride, and the apoprotein (which has no activity) is structurally much less stable to any of the dissociating treatments mentioned above.[79,94,95,97]

These stability properties should not be taken carte blanche as invariant characteristics of copper-zinc dismutases. First, the bovine enzyme itself becomes less stable on prolonged (months) frozen storage (with respect to ability to be dissociated by sodium dodecyl sulfate) even though its activity remains unimpaired.[74,79] Secondly, considerable variation is found among Cu-Zn enzymes from other species, even though they may be quite stable to the McCord-Fridovich protocol. For example, the enzyme from bakers' yeast[14] and wheat germ isozyme I[19] are dissociated by sodium dodecyl sulfate in the absence of mercaptoethanol. The yeast enzyme is more sensitive to denaturation and inactivation by 8 *M* urea than bovine erythrocyte dismutase.[98] Electrophoretic heterogeneity is seen in the *N. crassa* enzyme, when concentrated protein solutions are frozen.[13]

In summary, while Cu-Zn dismutases are a remarkably stable enzyme family, with respect to their stability to the McCord-Fridovich protocol, it may be anticipated that some species variation from the paradigm of the bovine erythrocyte protein may be encountered. In appreciation of this variability, it may be prudent to avoid indiscriminate exposure to extremes of denaturing or dissociating agents, to heat, or pH.

2. *Manganese/Iron Dismutases*

While the stability of Mn/Fe enzymes has not been studied as systematically as that of the dismutase from bovine erythrocytes, as a family they are noticeably less stable than Cu-Zn enzymes, but by no means are they "labile" enzymes. For example, the manganese dismutase from chicken liver is inactivated by chloroform-ethanol treatment,[26] and all reported procedures for purification of manganese- and iron-containing dismutases have used traditional procedures for cell breakage and protein fractionation; any exposure to organic solvents and phase separations by salting out have been conspicuously absent. In estimations of subunit molecular weight by sodium dodecyl sulfate-polyacrylamide gel electrophoresis, dissociation into constituent chains seems to proceed smoothly, without recourse to excessive heating or the presence of metal chelators.[26,37-39,40,41,45,48,51,54,56,58,61,62,64,66,69,71]

Compared to Cu-Zn enzymes, the activity of Mn/Fe dismutases is not as stable to alkaline pH.[38,40,99,100] In the case of manganese-containing dismutases, the difference in activity between pH 7.8 and 10.0 has been used to distinguish between Mn and Cu-Zn enzymes in crude homogenates.[74] Beyond these generalizations, there are variations between species. For example, the manganese dismutase from rat liver seems to be more labile to mildly acidic pH and to frozen storage[71] than the human liver enzyme.[74] Dismutases from thermophilic bacteria are more stable than their manganese-containing counterparts from mesophiles. For example, *Thermus aquaticus* dismutase is stable for hours in 8 *M* urea and in 6 *M* guanidine hydrochloride, 1 m*M* EDTA.[62] The apo protein of *Bacillus stearothermophilus* dismutase has been shown to be more stable than that of *E. coli* B, in metal reconstitution studies.[43,101-104]

D. Phylogenetic Distribution

1. *Introduction*

The isolation of three different SODs naturally led to the question of the distribution of manganese-, iron-, and copper-and-zinc-containing forms among living species. Distribution surveys are a necessary first step in investigating the fascinating questions of comparative biochemistry and evolution: why different organisms possess different metalloenzyme forms, what the adaptive significance of the specific forms might be, and whether there is one ancestral protein from which all three forms evolved.

An increasing number of SODs have been purified, and characterized as homogeneous proteins (Tables 1 and 2). However, most distribution studies have been performed using methods which allow the different forms to be distinguished in crude homogenates. Some of these methods are based upon differential sensitivity to inhibitors. Any given method for assaying dismutase activity is applicable to any of the three metalloenzyme forms, but by performing the assay in the presence of cyanide anion[74] or azide anion,[105] or by pretreatment of the crude extract with hydrogen peroxide,[56] it is possible (using the proper combinations) to infer which form(s) are present. Another method involves depleting the proteins in the extract of metals, and observing which metal(s) are most effective in reconstituting dismutase activity.[42] Still another approach, which is most applicable to microorganisms, is growth in the presence of a radioisotope of a metal ion, and assay for radioactivity which comigrates with dismutase activity after polyacrylamide gel electrophoresis.[106,107] These methods are discussed in more detail in the following chapter of this volume; their relative merits have been evaluated previously.[107]

2. General Distribution Pattern

Prokaryotes, both bacteria and blue-green algae, contain only enzymes of the Mn/Fe family. Some species contain only a single type, i.e., a manganese or an iron dismutase, while other species contain both metalloenzyme forms. Although exceptions exist, Gram-positive bacteria usually contain only a manganese dismutase, while Gram-negative bacteria generally contain only the Fe form or both forms.[2,60] A more reliable correlation has been made between metal co-factor and oxygen content of the growth environment: (1) anaerobes, anaerobically grown-facultative anaerobes, and aerobic diazotrophs contain only the iron-containing dismutase, while (2) species grown at higher oxygen concentrations contain only the manganese forms or both.[2]

The distribution of the three metalloenzyme forms among eukaryotes is more complex and shows an interesting dependency upon evolutionary pecking order.[2] The copper-zinc forms are found in all higher eukaryotes, from fungi, ferns, mosses and land plants, to invertebrates and vertebrates, while protozoans, the most primitive eukaryotes, lack Cu-Zn superoxide dismutases. The first appearance of the Cu-Zn form has been pinpointed to within the eukaryote green algae, of the class *Chlorophyceae.*[108] Chlorophycean algae with a phycoplast type of cell division (e.g., genera *Ulothrix, Ulva, Scenedesmus, Cladophora,* and *Codium*) lack Cu-Zn dismutases, while those with a phragmoplast type of cell division (eg., genera *Chara, Nitella,* and *Spirogyra*) and thus more closely allied with higher plants, do have Cu-Zn dismutases.

Manganese- and iron-containing SODs are found throughout the eukaryote realm, both in species which contain the Cu-Zn enzyme and those which lack it.[2] In several instances where both Cu-Zn and Mn forms are present (eg., yeast,[68] bovine heart,[72] chicken,[26] rat,[71] and human liver[73,74]) the Cu-Zn dismutase has been shown to be cytosolic and the Mn form to be associated with the mitochondria. Among these "mitochondrial" forms, there are great differences in the degree to which the manganese dismutase exclusively resides in that organelle. For example, the manganese enzymes from chicken liver and rat liver appear to be located predominantly in the mitochondria, while those from baboon (*Papio ursinus*) liver and from bakers' yeast appear to be found in both mitochondria and the cytosol fraction. Also, it may be pointed out that the quantity of manganese dismutase present in the liver of different species varies tremendously, from very low levels in rat liver to quite high levels in human and other primates.[73] The yeast mitochondrial enzyme has been shown to be encoded in the nucleus.[109]

The purification of an iron-containing dismutase from a higher plant (*Brassica campestris,* mustard leaves) was reported recently.[69] Among eukaryotes, the iron form had previously been found only in primitive eukaryotes, e.g., *Euglena gracilis,* which lack the Cu-Zn enzyme. This fascinating discovery raises the prospect that the iron form may be more widely distributed among higher eukaryote species, although there appear

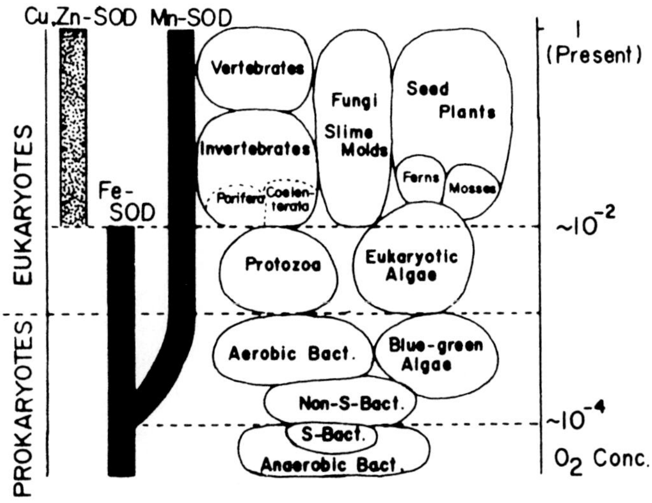

FIGURE 1. Phylogenetic distribution of Cu-Zn, Mn, and Fe SODs in organisms at different evolutionary levels. The scale at the right indicates the concentration of atmospheric oxygen, relative to the present atmospheric concentration, which is believed to have been present when the designated organisms appeared on earth. (From Asada, K., Kanematsu, S., Osaka, S., and Hayakawa, T., in *Chemical and Biochemical Aspects of Superoxide Dismutase*, Bannister, J. V. and Hill, H. A. O., Eds., Elsevier/North-Holland, New York, 1980. With permission.)

to be no reports of its presence in animal species. A summary of the known distribution of dismutase forms is shown schematically in Figure 1.

3. Exceptions

In 1974, a copper-and-zinc-containing SOD, termed bacteriocuprein, was discovered in the luminescent marine bacterium, *Photobacterium leiognathi.*[10] Bacteriocuprein is neither a typical bacterial dismutase, with atypical copper and zinc co-factors, nor is it a fourth form of SOD. Its properties clearly indicate that it is a bona fide member of the well-characterized family of Cu-Zn enzymes which had hitherto been found only in eukaryotes. Thus, by thermal stability, inactivation by 2,3-butanedione, molecular weight, and discriminant function analysis of its amino acid composition, it is phenotypically a copper-zinc SOD.[10,11] Initially two nonidentical subunits, of 17,000 and 15,500 mol wt, were reported for bacteriocuprein;[10] however a more recent preparation is a homodimer of 16,000 mol wt polypeptide chains, like all other Cu-Zn dismutases.[11]

The strain of *P. leiognathi* from which the bacteriocuprein was isolated occurs in nature as a symbiont in leiognathid fish (pony fish). These fish, which live in the western Pacific ocean, contain the bacterium in a light organ, which encircles the esophagus. By a system of apertures and reflectors, the bacterial luminescence can be used to illuminate the ventral surface of the pony fish, presumably to camouflage it against predation from below. The luminescence matches the sky light, which penetrates the ocean water, thus obscuring the silhouette of the fish.[110]

The long-standing symbiotic pact between fish and bacterium, plus the similarities in properties between bacteriocuprein and Cu-Zn dismutases of teleost fish and other vertebrates, has led to the suggestion that the bacterial gene for the dismutase may have been acquired by natural gene transfer from the pony fish host.[11] Natural gene transfer from prokaryote to eukaryote has been well documented in the crown gall tumor syndrome.[111,112] The eukaryote-to-prokaryote transfer, proposed as the origin of bacteriocuprein, is an intriguing hypothesis, which awaits further study.

A report recently appeared, describing a Cu-Zn dismutase in *Paracoccus denitrificans*, which may represent a second instance of a typically eukaryotic form of the enzyme residing in a prokaryote.[9] While *P. denitrificans* is not a symbiont, it has been pointed out that the hypothesis of gene transfer may be applicable here also: the bacterium resides in the soil and may have acquired the eukaryote gene as an enterobacterium of soil organisms, such as earthworms and nematodes.[113]

Do bacteriocupreins enjoy a distribution wider than the two known species? Some 93 strains, representing 23 species and biotypes of *Photobacterium*, and the related genera *Beneckea* and *Vibrio*, were screened by staining electrophoretograms of cell-free extracts for dismutase activity.[114] All extracts contained an iron-containing dismutase (based on inactivation by hydrogen peroxide), but only *P. leiognathi* contained a second dismutase, whose mobility agrees with that of the bacteriocuprein.[11] Lacking a compelling explanation for exclusive proprietary rights in *P. leiognathi* and *P. denitrificans* species, one wonders if the apparent absence of bacteriocupreins elsewhere is an experimental oversight. A bacteriocuprein might have been overlooked if it had an electrophoretic mobility identical to the iron form, particularly if present in much smaller amounts, or if its pI were so high that it did not penetrate the acrylamide gel (usually run at pH 8 to 8.5). Note that the cuprein of *P. leiognathi* has a pI of 7.9 to 8.4 (isozyme forms) compared to a value of about 4.4 for the iron-containing dismutase from the same organism.[11] The absence of copper-zinc dismutases in certain well-characterized bacterial species, such as *E. coli*, is well established. However, more rigorous examination of other species seems warranted, particularly other *Paracoccus* species, in light of the discovery of the *P. denitrificans* copper-zinc dismutase.

The luminous fungus, *Pleurotus olearius*, appears to be another exception to the general dismutase phylogeny. Two SODs have been isolated from this organism, both being manganese-containing forms. The apparent absence of a copper-zinc dismutase is surprising, in view of isolations of Cu-Zn enzymes from other fungi, *N. crassa*,[12,13] *F. oxysporum*,[12] and *S. cerevisiae*.[14-16] It would seem to be important to examine homogenates of *P. olearius* for the presence of a cyanide-sensitive dismutase activity.

4. The Existence of SOD in Anaerobes; Adaptive Significance of Dismutase Forms

Shortly after the discovery of SOD, levels of its activity were surveyed in microorganisms of aerobic, facultatively anaerobic, and obligately aerobic persuasions.[115] SOD activity was not found in obligate anaerobes, while it was present in the other microbes. This evidence was interpreted in favor of an "enzyme-based theory of aerobiosis": that SOD protected aerobes against the toxic by-products of oxygen metabolism and that its absence from strict anaerobes was consistent with the inability of anaerobes to survive in oxygen-containing environments. Since that time, SOD activity has not only been documented in obligately anaerobic organisms, but the enzyme (an iron form in all cases) has been purified from several:[2] from *Chromatium vinosum*,[38] a purple sulfur bacterium, from *Chlorobium thiosulfatophilum*,[37] a green sulfur bacterium, and from the sulfate reducing bacterium, *Desulfovibrio desulfuricans*.[39] Oxygen toxicity studies in prokaryotes and eukaryotes (described in Chapters 4 and 5 of this volume) have provided more compelling evidence than the above distribution survey, for the physiological role of SOD in scavenging the superoxide anion. Nonetheless, the existence of SOD in anaerobes has remained a meaty bone of contention for those who are skeptical about the physiological clout of the enzyme as a scavenger of superoxide. How can the presence of an SOD be rationalized in organisms which live in anaerobic niches? Several possible replies have been elaborated.[2,38]

One reply is that anaerobic SOD is a microbial appendix, that it is of little current physiological significance, either a protein retained from a previous period when the microorganism lived aerobically, or else a protein acquired by a recent gene transfer, perhaps as an adventitous rider on a plasmid package,[39] which principally delivered genes for other, more needed functions. There are, however, several observations which suggest a current physiological function. (1) The anaerobe, *Chromatium*, can survive for a short time, without growth, in aerobic conditions, suggesting that dismutase is utilized for scavenging the superoxide radical during such brief forays into an oxygen-containing environment. (2) Even when grown under strictly anaerobic conditions, anaerobes contain an iron SOD (the only form found in anaerobes) at levels comparable to dismutases in aerobes. Likewise, the iron form is the only dismutase present when facultative aerobes, such as *E. coli*, are grown anaerobically. (When grown aerobically, these organisms generally contain both iron and manganese dismutases.) These data suggest that the iron dismutase has been retained in obligate anaerobes and anaerobically grown facultative anaerobes for a function in the absence of oxygen.[2,38] Such a suggestion does not deny a physiological role for dismutase in scavenging the superoxide radical, but rather raises the possibility (which is sheerly speculative at present) that the enzyme may have additional functions.[2,6,38]

SOD can inhibit the reaction of semiquinones with methemoglobin, presumably by effecting conversion of the reactive semiquinone into the unreactive quinone by catalyzing decomposition of superoxide, $O_2^{\overline{\cdot}}$, in the following equation.[116]

$$\text{Semiquinone} + O_2 \rightleftharpoons \text{quinone} + O_2^{\overline{\cdot}} \qquad (1)$$

While this particular feat of inhibiting reactions of semiquinone free radicals is merely a consequence of $O_2^{\overline{\cdot}}$ dismutation and is not a new activity of SOD, it is perhaps illustrative of the "oblique thinking"[117] which may yield new insights into physiological functions of this enzyme.

The enigma of SOD in anaerobes is but one facet of the more pervasive question of the adaptive significance a metalloenzyme form has for the particular organism in which it resides. One matter is perfectly clear: that the metalloenzyme form is not a consequence of habitat; organisms in seemingly identical environments may contain different dismutase forms.[2] The answer to this question clearly lies in the physiological idiosyncrasies of oxygen metabolism in different species.

The identical question of adaptive significance may be asked about the bacteriocupreins. What peculiar physiological traits of *Photobacterium leiognathi* and *Paracoccus denitrificans* require the presence of a eukaryote, copper-and-zinc-based SOD? The hypothesis of a plasmid origin to the bacterial gene for bacteriocuprein is a particularly attractive one,[11] in view of the analogies between the protective function of SOD and that of antibiotic resistance elements, which are known to exist on plasmids, such as the R factor.[118] If this hypothesis is true, then what selection pressures have favored retention of the plasmid in the bacterium?

E. Evolutionary Relationships and Amino Acid Sequences

1. Introduction

The segregation of dismutases into two families, and the similarities of manganese- and iron-containing forms, are amply documented by purification properties and stability behavior. However, questions of whether corresponding homologies and disparities appear in the protein structures themselves, as well as questions of evolutionary relatedness, require more rigorous chemical characterizations. This section describes the result of comparative amino terminal sequence analyses of SOD and its implications for the uniqueness of the evolutionary conundrum among these enzymes. The

Species and metal cofactor	Amino terminal sequence (residues 1–18)
Obligate anaerobes	
Chromatium vinosum, iron	Met-His-Glu-Leu-Pro-Ala-Leu-Pro-Tyr-Glu-Lys-Asn-Ala-Leu-Glu-Pro-Val-Ile
Chlorobium thiosulphatophilum, iron	Ala-Tyr-Glx-Glx-Pro-Ala-Leu-Pro-Tyr-Ala-Asx-Asx-Ala-Leu-Glx-Pro-His-Ile
Desulphovibrio desulphuricans, iron	Ser-Ile-Phe-Val-Leu-Pro-Asp-Leu-Pro-Tyr-Ala-Lys-Asp-Ala-Leu- ? -Pro-Lys-Ile
Facultative and obligate aerobes and algae	
Bacillus stearothermophilus, manganese	Pro-Phe-Glu-Leu-Pro-Ala-Leu-Pro-Tyr-Pro-Tyr-Asp-Ala-Leu-Glu-Pro-His-Ile
Escherichia coli B, manganese	Ser-Tyr-Thr-Leu-Pro-Ser-Leu-Pro-Tyr-Ala-Tyr-Asp-Ala-Leu-Glu-Pro-His-Phe
Escherichia coli B, iron	Ser-Phe-Glu-Leu-Pro-Ala-Leu-Pro-Tyr-Ala-Lys-Asp-Ala-Leu-Ala-Pro-His-Ile
Photobacterium leiognathi, iron	Ala-Phe-Glu-Leu-Pro-Ala-Leu-Pro-Phe-Ala-Met-Asn-Ala-Leu-Glu-Pro-His-Ile
Plectonema boryanum, iron	Ala-Tyr-Thr-Gln-Pro-Pro-Leu-Pro-Phe-Pro-Lys-Asp-Ala-Leu-Glu-Pro-Tyr- ?
Pseudomonas ovalis, iron	Ala-Phe-Glu-Leu-Pro-Pro-Leu-Pro-Tyr-Ala-His-Asp-Ala-Leu-Gln-Pro-His-Ile
Rhodopseudomonas sphaeroides, manganese	Ala-Phe-Thr-Leu-Pro-Asp-Leu-Pro-Tyr-Ala-His-Asp-Ala-Leu-Ala-Ala-Leu-Gly
Spirulina platensis, iron	Ala-Phe-Glu-Leu-Pro-Ser-Leu-Pro-Phe-Asp-Gln-Asp-Ala-Leu-Glu-Ser-Ser-Lys
Thermus aquaticus, manganese	Pro-Tyr-Pro-Phe-Lys-Leu-Pro-Glu-Leu-Gly-Tyr-Pro-Tyr-Glu-Ala-Leu-Glu-Pro-His-Ile
Eukaryotes	
Saccharomyces cerevisiae, manganese	Lys-Val-Thr-Leu-Pro-Asp-Leu-Lys-Trp-Asp-Phe-Gly-Ala-Leu-Glu-Pro-Tyr-Ile
chicken liver, manganese	Lys-His-Thr-Leu-Pro-Asp-Leu-Pro-Tyr-Asp-Tyr-Gly-Ala-Leu-Glu-Pro-His-Ile
human liver, manganese	Lys-His-Ser-Leu-Pro-Asp-Leu-Pro-Tyr-Asp-Tyr-Gly-Ala-Leu-Glu-Pro-His-Ile
Saccharomyces cerevisiae, copper-zinc	Val-Gln-Ala-Val-Ala-Val-Leu-Lys-Gly-Asp-Ala-Gly-Val-Ser-Gly-Val-Val
horse liver, copper-zinc	Ac-Ala-Leu-Lys-Ala-Val-Cys-Val-Leu-Lys-Gly-Asp-Gly-Pro-Val-His-Gly-Val-Ile
bovine erythrocytes, copper-zinc	Ac-Ala-Thr-Lys-Ala-Val-Cys-Val-Leu-Lys-Gly-Asp-Gly-Pro-Val-Gln-Gly-Thr-Ile
human erythrocytes, copper zinc	Ac-Ala-Thr-Lys-Ala-Val-Cys-Val-Leu-Lys-Gly-Asp-Gly-Pro-Val-Gln-Gly-Ser-Ile

Species and metal cofactor	Sequence (residues 19–37)
Obligate anaerobes	
Chromatium vinosum, iron	-Ser-Ala-Glu-Thr-Ile-Glu-Tyr-His-Tyr-Gly- ? - ? -His-Gln-Thr-Tyr-Val-Thr-Asn-
Chlorobium thiosulphatophilum, iron	- ? -Ala-Glx-Thr-Ile-Gly-Phe-His-Tyr-Gly-Lys-His-His-Ala-Ala-Tyr-Val-Lys-Thr-
Desulphovibrio desulphuricans, iron	-Ser-Ala-Lys-Thr-Phe-Asp-Phe- ? - ? -Gly-Lys-
Facultative and obligate aerobes and algae	
Bacillus stearothermophilus, manganese	-Asp-Lys-Glu-Thr-Met-Asn-Ile-His-His-Thr-Lys-His-His-Asn-Thr-Tyr-Val-Thr-Asn-
Escherichia coli B, manganese	-Asp-Lys-Gln-Thr-Met-Glu-Ile-His-His-Thr-Lys-His-His-Gln-Thr-Tyr-Val-Asn-Asn-
Escherichia coli B, iron	-Ser-Ala-Glu- ? -Ile-Glu-Tyr-His-Tyr-Gly-Lys-
Photobacterium leiognathi, iron	-Ser-Gln-Glu-Thr-Leu-Glx-Tyr-Gly-Tyr-Gly-Lys-His-His-Asx-Thr-Tyr-Val- ? -Lys-
Plectonema boryanum, iron	-Met- ? -Ala-Glu- ? -Phe-
Pseudomonas ovalis, iron	-Ser-Lys-Glu-Thr-Leu-Glu-Tyr-His-His-Asx-Lys-His-His-Asn-Thr-Tyr-Val-Val-Asn-
Rhodopseudomonas sphaeroides, manganese	-Met-Met-Lys-Glu-Thr-Met-Glu-Tyr-His-His-Asp-Ile-His-His-Lys-Ala-Tyr-Val-Asp-Asn-
Spirulina platensis, iron	-Met-Ser-Ala-Asn-Thr-Leu-Ser-Tyr-His-His-Gly-Lys-His-His-Ala-Ala-Tyr-Val-Lys-Asn-
Thermus aquaticus, manganese	-Asp-Ala-Arg-Thr-Met-Glu-Ile-His-His-Gln-Lys-His-His-Gly-Ala-Tyr-Val-Thr-Asn-
Eukaryotes	
Saccharomyces cerevisiae, manganese	-Ser-Gly-Gln-Ile-Asn-Glu-Leu-His-Tyr-Thr-
chicken liver, manganese	-Ser-Ala-Glu-Ile-Met-Gln-Leu-His- ? - ? -Lys-
human liver, manganese	-Asn-Ala-Gln-Ile-Met-Gln-Leu-His- ? -Ser-Lys-
Saccharomyces cerevisiae, copper-zinc	-Lys-Phe-Glu-Gln-Ala-Ser-Glu-Ser-Glu-Pro-Thr-Thr-Val-Ser-Tyr-Glu-Ile-Ala-Gly-Asn-
horse liver, copper-zinc	-His-Phe-Glu-Gln-Gln-Gln-Glu-Gly-Gly-Pro-Val-Val-Leu-Lys-Gly-Phe-Ile-Glu-Gly- -
bovine erythrocytes, copper-zinc	-His-Phe-Glu-Ala-Lys-Gly- - -Asp-Thr-Val-Val-Val-Thr-Gly-Ser-Ile-Thr-Gly- -
human erythrocytes, copper-zinc	-Asn-Phe-Glu-Gln-Lys-Glu-Ser-Asp-Gly-Pro-Val-Lys-Val-Trp-Gly-Ser-Ile-Lys-Gly- -

FIGURE 2. Amino terminal sequences of manganese-, iron-, and copper-zinc-containing SODs. Within each family (Mn/Fe or Cu-Zn) the residues which are identical in the sequences from all species are enclosed in boxes.[77-79,81-84,121,122]

results of complete amino acid sequence determinations (presented in Sections III.D and IV.B) confirm the conclusions drawn from amino terminal segments, although no complete sequence data have been reported for any iron-containing dismutase, or for a manganese-containing dismutase from a eukaryote (ie., a "mitochondrial" enzyme).

Evolutionary relationships, traced in a great number of protein families through comparative amino acid sequence analysis, have established the following principles. (1) Two proteins which have evolved from a common ancestral protein (divergent evolution) will share regions of extensive sequence identity, or conservative amino acid substitutions at corresponding sequence positions. The greater the time since divergence, the less will be the homology between the sequences. (2) During the same period of time, different protein families will have diverged at different rates, and the degree of homology between two different species will depend upon which protein family is used for the comparison. (3) In a comparison of two protein sequences, the degree of homology will generally not be constant over the polypeptide chain, but will be higher in regions which are essential for the biological function or structural integrity of that family.[119,120]

2. Results and Conclusions from Sequence Comparisons

The amino terminal sequences of 15 dismutases from the Mn/Fe family and 4 dismutases from the Cu-Zn family (shown in Figure 2[77-84,121,122] are aligned to maximize the homology within each family. The immediate conclusion from Figure 2 is that there is a high degree of homology among sequences of the same family consistent with the similarities in purification and stability properties. Within the Mn/Fe family 8 of the first 40 residues are absolutely invariant, and other positions have only one or two exceptions to invariance.[121] Such invariance over the astounding phylogenetic range from man to microorganisms suggests that Mn and Fe forms evolved from a common ancestor. In the Cu-Zn family, 12 of the first 40 residues are invariant over an equally vast phylogenetic range, and an identity of nearly 50% is found when the entire amino acid sequences are compared.[80] This indicates that the members of the Cu-Zn family have diverged very slowly, because their sequence conservation is almost as great as that of the eukaryotic cytochromes *c*.[80] Such conservation is considered characteristic of proteins with a crucial physiological role, and is consistent with the protective function proposed for dismutases.

The lack of similarity between the Cu-Zn and the Mn/Fe families is as apparent as the similarity within each. Whether deduced by visual inspection or computer analysis, whether based solely upon amino terminal regions or upon the available complete amino acid sequences, the conclusion is the same.[44,89,121,122] The lack of significant sequence homology between Mn/Fe and Cu-Zn families and the gross differences between the three-dimensional structure known for the bovine Cu-Zn enzyme[85-87] and the secondary structure predicted from the two known sequences of manganese-containing dismutases,[44,122] strongly suggest that these two families arose from two different ancestral proteins. Clearly, Mn/Fe and Cu-Zn families are not the product of divergent evolution from a single ancestor. SODs thus represent an evolutionary pattern different from that of the α and β chains of hemoglobin, which evolved from a common, myoglobin-like precursor.[119]

Within the Mn/Fe family, it is important to note that a sequence homologies between anaerobes and aerobes are just as great as those between one aerobic species and another. Thus the iron-containing dismutases found in anaerobes are not a subclass unto themselves and are not significantly different from any other Mn/Fe dismutase in their amino terminal sequences.[121,122] It is reasonable to propose then that the common ancestor of present day Mn and Fe dismutases was an iron-containing form, because anaerobes contain exclusively the Fe form and because anaerobic organisms are believed to have preceded aerobic species in the evolution of life.[2] It has been proposed that the histidine residues at positions 27, 31, and 32, may be functionally important, perhaps as metal ligands, based upon their invariance or near invariance in the known amino terminal sequences of Mn/Fe enzymes.[121]

3. Significance of Dismutase Evolutionary Relationships

The sequence homologies within the Mn/Fe family include extensive similarities between the dismutases from prokaryotes and the "mitochondrial" dismutases. Bacterial-mitochondrial similarities, whether evidences in ultrastructure, protein synthesis mechanisms, or in amino acid sequences, have been taken as evidence for the endosymbiotic hypothesis of mitochondrial origin, ie., that mitochondria evolved from primitive prokaryotes, which lived symbiotically in proto eukaryotes.[123] The endosymbiotic relevance of bacterial-mitochondrial similarities is often disparaged, by ascribing the similarities to a trait dating from before the divergence of eukaryote and prokaryote cell lines and simply sustained in both lines.[124] However, this argument of common prokaryote-eukaryote ancestry is not as strong when leveled against the SODs. The earth initially contained a reducing atmosphere. It is reasonable that SOD, acting on an oxygen by-product, would have evolved only after the accumulation of atmospheric oxygen was sufficient to constitute an oxygen toxicity threat. Oxygen accumulation is attributed to the photosynthetic production of oxygen by blue-green algae, and these algae are dated near the time of divergence of the eukaryotes and prokaryotes. Consequently, the first dismutase enzymes probably occurred after the divergence of the two cell lines, and any similarity between eukaryote and bacterial dismutases would be more consistent with an endosymbiotic event than with an inherited trait.[89,122,123]

While sequence data establish that the two dismutase families have not evolved from a common ancestor, such data do not distinguish whether the two families evolved in parallel fashion (ie., each evolving its own characteristic structure and catalytic mechanism) or whether in convergent fashion (i.e., evolving a similar catalytic mechanism, housed in a very different protein structure.)[119] The proteases, trypsin and subtilisin BPN′, are an example of convergent evolution, because they have appropriated highly homologous charge relay mechanisms of peptide bond hydrolysis within very different protein fabrics.[125] The distinction between parallel and convergent evolution will probably require identification of the metal ligands and information about the ligand geometry, from X-ray crystallographic and spectroscopic analyses of manganese- and iron-containing dismutases.

Perhaps the most fascinating and the most unique feature of dismutase evolution is the invention of a totally new form of the enzyme, the Cu-Zn form of eukaryotes, when two catalytically competent forms, the Mn and Fe forms, had been in existence since the emergence of aerobic prokaryotes. It may be presumed that when Cu-Zn dismutases emerged, during the speciation of eukaryote green algae,[108] there existed a particularly compelling selection pressure, which necessitated a new type of SOD of radically different sequence and structure and with different metal co-factors. Furthermore, the physiology of oxygen metabolism in higher eukaryotes must remain, to this day, sufficiently different from that of prokaryotes to favor retention of the gene for Cu-Zn SOD. Little is known about such selection pressures and said physiological distinctions, and again the specter, that dismutases may engage in meaningful activities other than the dismutation of superoxide free radical, might be raised.

Eukaryotes contain a Cu-Zn SOD predominantly in the cytosol and an Mn/Fe dismutase in the mitochondria and/or cytosol. Eukaryote SODs are thus another example of a protein which exists in two discrete forms within the same cell type. This phenomenon is not unusual and has been documented in other protein families including the iso-cytochromes *c* in yeast,[126] the ferredoxins,[127,128] and the aspartate aminotransferases.[129-131] However, in these three other instances, the amino acid sequences of the two protein forms are very similar, differing in only a small percentage of the sequence positions. This small sequence difference is evidently sufficient for producing the necessary differences, in functional, or regulatory, or other physiological properties, between the two forms. The Cu-Zn and Mn/Fe SODs on the other hand, show no significant sequence homology. One wonders what dramatically different physiological roles necessitated such different structures. The difference does not seem to be a necessity for transporting the manganese form into the mitochondrion, because the mitochondrial aspartate aminotransferase incurs the same need, yet it is highly homologous to the cytoplasmic form of the enzyme.[122,129-131]

It is rather intriguing to note that the diversity seen between the two families of SODs is in fact found in other families of proteins which deal with oxygen or oxygen reduction products. Among peroxidases, for example, there are catalases based upon heme[132] and there is glutathione peroxidase, which is a selenium-based enzyme.[133] Among oxygen transport proteins, in vertebrates, there are heme-based hemoglobins and myoglobins, and in invertebrates, blue copper-based hemocyanins and nonheme iron-based hemerythrins. It is not particularly surprising that the profound selection pressures, accompanying the oxygenation of the atmosphere, evoked specific proteins for transporting oxygen and scavenging its toxic by-products. What is surprising is that these proteins exhibit a diversity in structure and mechanism of action which seems to surpass that found in other families of proteins whose members share a common function.

III. COPPER-ZINC SODs

A. Introduction

This section deals with structure-function relationships: how the catalytic mechanism and structural stability of Cu-Zn dismutases is understood in terms of their overall molecular conformation and the contributions of specific amino acid side chains. The three-dimensional structure of the Cu-Zn enzyme from bovine erythrocytes (the only dismutase from either family for which a crystal structure has been solved) is a logical starting point, because it has been a focal point for these structure-function studies. While wielding considerable clout in this arena, the crystal structure has not always been the ultimate fount of knowledge. Spectroscopic and chemical studies have provided confirmation and interpretation and have generated new insights about structure and mechanism. By necessity, emphasis in this section will be upon the Cu-Zn enzyme from bovine erythrocytes, because it has been the object of most of these studies. However, the validity of structure-function correlations which are deduced for this single protein, as well as an appreciation of the nuances which may reveal the molecular basis of physiological adaptation, require interspecies comparisons. For this reason, comparative data for Cu-Zn dismutases from other species will be presented, when available.

B. The Three-Dimensional Structure of Crystalline Copper-Zinc Dismutase from Bovine Erythrocytes

1. Molecular Conformation and Structure of the Catalytic Site

The three-dimensional structure has been determined to 3Å resolution[85,86] and is being refined to a resolution of 2Å.[134] The two subunits of the dimer, known to be chemically identical,[77-79] appear to have identical, or nearly identical, backbone conformations at 3Å resolution. The minor differences which can be seen are probably attributable to nonequivalent crystal packing environments. The backbone conforma-

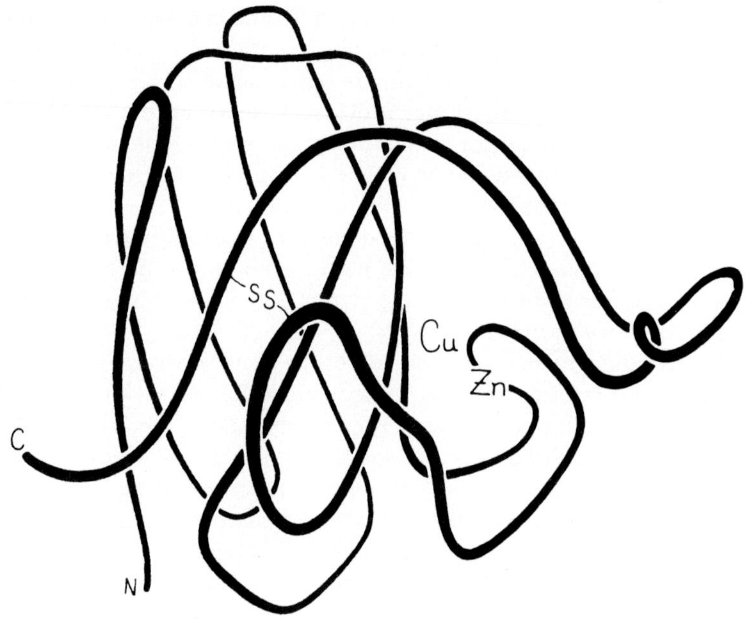

A

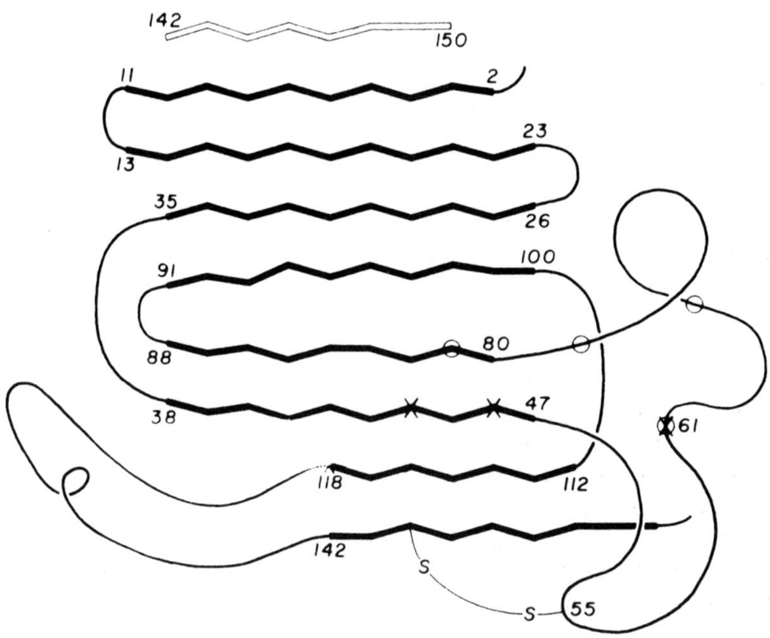

B

FIGURE 3. Representation of the polypeptide backbone structure of the subunit of bovine erythrocyte copper-zinc SOD. (A) Schematic drawing of the three-dimensional structure. (B) Two-dimensional representation of the topological relationship of the eight strands of β-sheet structure. The carboxyl terminal strand (residues 142 to 150) is repeated at the top of the figure, to demonstrate its position relative to the amino terminal strand. Residues which contribute ligands to zinc are designated by circles; those which ligand to copper are indicated by Xs. (With permission from Richardson, D. C., in *Superoxide and Superoxide Dismutases*, Michelson, A. M., McCord, J. M., and Fridovich, I., Eds., Academic Press, London, 1977, pp. 218 and 220. Copyright by Academic Press, Inc. [London], Ltd.)

tion of a single subunit is shown in three- and two-dimensional representations, in Figures 3A and 3B, respectively.[86]

The predominant feature of the structure is its large amount of β-sheet. Best seen in Figure 3B, this sheet contains nearly 50% of the sequence, in 8 strands of antiparallel arrangement. In three dimensions, the strands comprise a cylinder or barrel. The subunit interface is at the front of the β-barrel (in Figure 3A), slightly left of the center. The contact between the subunits is principally made by three polypeptide segments. Two of these are segments joined by the intrachain disulfide bond; the third is a short segment, which appears at the bottom center of Figure 3A, bending towards the back of the subunit. At the subunit interface, and in many of the turns between β-strands, there reside glycine residues which are probably crucial for closeness of approach and sharp bending, respectively.

The β-barrel structure in the enzyme led the dismutase crystallographers to notice that repeating β-sheet structures were not at all uncommon in proteins. A classification has been published of the folding patterns observed in proteins with such β-barrel secondary structures.[135] Are proteins with topological similarity in their β-sheet structures evolutionarily related? The answer appears to be "no". The topological similarity evidently arises from thermodynamic considerations, i.e., an energetic selection rather than a neo-Darwinian natural selection. The energetics evidently highly favor only a very few among the many possible polypeptide folding patterns.[135]

The residues which are not in the β-sheet structure of bovine Cu-Zn dismutase comprise two long loops of nonrepetitive structure, which lie on the surface and tend to extend out from the molecule. (Both loops appear on the right in Figure 3A; in Figure 3B, one is at either side.) One loop (at the right in Figure 3B) contains several metal ligands and half-cystine 55 of the disulfide bond. The second loop (at the left in Figure 3B) contains no metal ligands or half-cystine, but does appear to have a unique conformation, although it extends rather far into solution and is stabilized by neither metal coordination nor disulfide bonding.

The metal-binding sites for copper and zinc are on the "back side" of the molecule, opposite the subunit interface (Figure 3A). The two copper ions on the two separate subunits are about 34 Å apart. The copper and zinc ions within each subunit are only about 6 Å distant, and are liganded by imidazole and carboxyl side-chain functions from residues in the β-sheet and loop structures of the molecule (Figure 3B).

The metal-binding sites have been subject to extensive scrutiny by spectroscopic techniques, and their depiction from the crystallographic analysis is therefore described in some detail. The four zinc ligands His 61, His 69, His 78, and Asp 81, are in approximately tetrahedral arrangement. The four ligands of copper, His 44, His 46, His 61, and His 118, define a distorted square planar arrangement, the distortion being that an imidazole nitrogen of His 61 does not lie in the same plane as the other 3 histidine ligands and the copper, but is located intermediate between planar and axial positions.[86,87] This distortion is proposed as the explanation for the rhombicity of the electron paramagnetic resonance spectrum of the enzyme, i.e., the deviation from the spectrum expected for copper in classic square-planar geometry.

The crystallographic data show that the imidazole of His 61 is situated between the copper and zinc ions and is close enough to interact with both metals. This imidazole has been proposed as a shared or bridging ligand, although the ring is not directly on a line between the Cu and the Zn. A large number of spectroscopic studies have focused upon this bridging imidazolate, its influence upon interactions between copper and zinc, its possible role in catalysis, and whether it is actually liganded to both metals (see Section III.G.5).

2. Relevance of the Structure to Stability Properties

One of the most worthwhile benefits of a crystal structure, which often continues to yield dividends after the crystallographic publication, is the explanations and correlations it offers for solution data about the protein. For example, the marked decrease in the stability of the protein upon reducing the intra-subunit disulfide bond[19,79] is easily understood when the subunit interface is seen to be formed in large part by the two polypeptide segments which are joined by the disulfide linkage (Figure 3A). The importance of the metal(s) for structural stabilization and the decreased stability shown by the apoprotein[79,94,97] are likewise plausible when it is seen that the binding site is relatively exposed and that some of the metal ligands are side chains in a loop, which extends out from the molecule.

Crystallographic data are often of limited applicability in predicting how a protein conformation will change in different solvent conditions or when metal co-factors are removed. Little can be said about structural changes (if any) which occur on partitioning of Cu-Zn dismutases into organic solvents (in the McCord-Fridovich purification scheme),[1] or to the maintenance of their activity in aqueous-organic mixtures.[136] Structural differences between the apo- and holoprotein have been analyzed by circular dichroism (CD) and proton nuclear magnetic resonance (NMR) spectroscopy. The β-sheet structure of the holoprotein appears to be present in the apo form, because the CD spectra of the two proteins are similar.[137] (There is one report of differences between the CD spectra of apo and holo bovine Cu-Zn dismutases.[138]) However, the apoprotein has a flexible, more dynamic structure than holoprotein, because greater resolution is seen in the proton NMR spectrum of apoprotein, due to averaging of anisotropic magnetic interactions.[97] The conclusion from these two studies is that metal(s) in the Cu-Zn dismutase serve to stabilize and organize secondary structural elements of the holoprotein, many of which already exist in the apo form.[139] The role of copper and zinc in the structure and mechanism of the active site have been extensively studied by spectroscopic methods (see Section III.G).

C. Chemical Modification of Bovine Dismutase and Other Copper-Zinc Dismutases

Histidine residues had been implicated as metal ligands by chemical modifications of bovine dismutase performed before the crystal structure was solved.[140,140a] The strategy in such studies has been to separately modify the holoenzyme and the metal-free apoprotein, with the same reagent, under the same reaction conditions. Residues which are reactive in the metal-free protein but which are unreactive in the holo form are therefore "protected" by the metal co-factors and presumed to be metal ligands. This approach for identification of ligands is fraught with complications. The principal assumption is that the backbone and side-chain conformations of apo and holo proteins are identical or nearly identical, and that the only significant difference between the two is the presence or absence of the metal ion. The principal complication is that it is usually not known whether removing the metal induces conformational changes (at the binding site or elsewhere), which change the accessibility and chemical reactivity of residues that may not be ligands.

Nonetheless, chemical modification of bovine dismutase, whether by dye-sensitized photooxidation, diazotized sulfanilamide,[140a] or by hydrogen peroxide,[140] yielded the same, correct answer: histidine residues as ligands. Although more flexible, the conformation of the apoprotein is evidently similar enough to that of holoprotein, that the only significant changes in reactivity occur in its metal-binding site residues. With less stable dismutases, enzymes from the Mn/Fe family or less hearty members of the Cu-Zn family, the changes in reactivity and conformation, associated with metal depletion, may be more significant, and the limitations of such a chemical modification approach for deducing metal ligands may be more severe.

Inactivation of bovine dismutase by hydrogen peroxide has been studied in greater detail, perhaps because H_2O_2 is a product of the dismutation reaction. The "resting"

Cu-Zn enzyme is in the cupric (Cu^{+2}) state.[141] Low concentrations of hydrogen peroxide merely reduce the metal from cupric to cuprous; higher concentrations cause irreversible inactivation.[140,142] Hydrogen peroxide is a rather nonspecific oxidizing and reducing agent, and its discriminating action on bovine dismutase has been attributed to the formation of a reactive oxygen species by complexation with the copper.[142] Why doesn't in vivo hydrogen peroxide, accumulated from dismutation of superoxide radicals, inactivate the enzyme? Evidently, the concentration of hydrogen peroxide is lowered in vivo by catalase and peroxidase activity. Also, inactivation studies were performed about pH 10; the rate of inactivation is much lower at physiological pH.[142]

A similar question may be asked about superoxide: if the superoxide radical is so reactive, then how can the enzyme survive repeated turnovers? Xanthine oxidase is autoinactivated by the superoxide radical which is generated during catalysis.[143] How does dismutase survive unscathed? Given that superoxide is an anion at physiological pH, an electrostatic guidance system, orienting the radical to the active site with great accuracy, seems a reasonable possibility. Recently, such a mechanism may have been found.[144-147]

Bovine erythrocyte dismutase is inactivated by butanedione. Arginine residues are selectively modified, to an extent of 1.0 to 1.3 residues per subunit, at 77 to 80% inactivation. Peptide mapping suggests that Arg 141 is the principal site of modification.[144,145] The guanidinium side chain of Arg 141 is known to form one side of the superoxide-binding pocket and to be within hydrogen-bonding distance of one imidazole nitrogen of His 46, whose other imidazole nitrogen is a copper ligand.[134] It has consequently been suggested that the Arg 141 side chain may function for electrostatic guidance of superoxide, or function as a proton donor in the second step of catalysis, the oxidation of cuprous to cupric with reduction of superoxide to peroxide (see Section III.F.1). Alternatively, the butanedione-modified side chain may simply perturb the active site because it is close to the copper atom, and Arg 141 may actually not be a catalytic participant itself.[144,145]

The specificity of α, β-diketones for modification of arginine residues lends these reagents to surveying other enzymes of the Cu-Zn family for functionally important guanidinium side chains. The Cu-Zn dismutase from bakers' yeast is inactivated by phenylglyoxal and also by a chromophoric analog, 4-hydroxy-3-nitrophenylglyoxal.[146,147] With these reagents, a loss of 1.0 to 1.04 residues of arginine per subunit corresponded to complete inactivation. Characterization of peptides from yeast enzyme, modified with the chromophoric analog, indicated reaction exclusively at Arg 143 in the sequence of the yeast enzyme,[147] which is homologous to bovine Arg 141.[82,83]

Analogous studies have been performed on dismutases from a number of species. Cu-Zn dismutases purified from other eukaryote sources (bovine liver, chicken liver, porcine liver, and wheat germ) were inactivated by phenylglyoxal, with activity losses of 84 to 94% corresponding to modification of 1.0 to 1.3 residues of arginine per subunit.[144,145] The bacteriocuprein from *Photobacterium leiognathi* and Cu-Zn dismutases purified from its pony fish host and from other teleost fishes were inactivated by butanedione, although modification of arginine was not demonstrated by amino acid analysis.[11] Enzymes of the Mn/Fe family show considerably less or no inactivation by butanedione.[144,145] In conclusion, the evidence is fairly convincing that an arginine residue, homologous to bovine Arg-141, has been conserved throughout evolution of Cu-Zn dismutases, that it lies in the metal-binding region, and that it may play a role in catalysis.

While all Cu-Zn dismutases contain phenylalanine, the amino acids tryptophan and tyrosine are scarce. Per subunit, the human erythrocyte enzyme contains one tryptophan[80,81] and the bovine enzyme one tyrosine.[77-79] These residues are not crucial for catalysis, because their modification (using *N*-bromosuccinimide and/or methylene

	1 … 10 … 20
bovine	Ac-Ala-Thr-Lys-Ala-Val-Cys-Val-Leu-Lys-Gly-Asp-Gly-Pro-Val-Gln-Gly-Thr-Ile-His-Phe-
human	Ac-Ala-Thr-Lys-Ala-Val-Cys-Val-Leu-Lys-Gly-Asp-Gly-Pro-Val-Gln-Gly-Ser-Ile-Asn-Phe-
horse	Ac-Ala-Leu-Lys-Ala-Val-Cys-Val-Leu-Lys-Gly-Asp-Gly-Pro-Val-His-Gly-Val-Ile-His-Phe-
yeast	Val-Gln-Ala-Val-Ala-Val-Leu-Lys-Gly-Asp-Ala-Gly-Val-Ser-Gly-Val-Val-Lys-Phe-

	21 … 30 … 37
bovine	Glu-Ala-Lys-Gly- - -Asp-Thr-Val-Val-Val-Thr-Gly-Ser-Ile-Thr-Gly- -Leu-Thr-
human	Glu-Gln-Lys-Glu-Ser-Asp-Gly-Pro-Val-Lys-Val-Trp-Gly-Ser-Ile-Lys-Gly- -Leu-Thr-
horse	Glu-Gln-Gln-Gln-Glu-Gly-Gly-Pro-Val-Val-Leu-Lys-Gly-Phe-Ile-Glu-Gly- -Leu-Thr-
yeast	Glu-Gln-Ala-Ser-Glu-Ser-Glu-Pro-Thr-Thr-Val-Ser-Tyr-Glu-Ile-Ala-Gly-Asn-Ser-Pro-

	40 … 50 … 57
bovine	Glu-Gly-Asp-His-Gly-Phe-His-Val-His-Gln-Phe-Gly-Asp-Asn-Thr-Gln-Gly-Cys-Thr-Ser-
human	Glu-Gly-Leu-His-Gly-Phe-His-Val-His-Gln-Phe-Gly-Asn-Asp-Thr-Ala-Gly-Cys-Thr-Ser-
horse	Lys-Gly-Asp-His-Gly-Phe-His-Val-His-Glu-Phe-Gly-Asp-Asn-Thr-Gln-Gly-Cys-Thr-Thr-
yeast	Asn-Ala-Glu-Arg-Gly-Phe-His-Ile-His-Glu-Phe-Gly-Asp-Ala-Thr-Asn-Gly-Cys-Val-Ser-

	60 … 70 … 77
bovine	Ala-Gly-Pro-His-Phe-Asn-Pro-Leu-Ser-Lys-Lys-His-Gly-Gly-Pro-Lys-Asp-Glu-Glu-Arg-
human	Ala-Gly-Pro-His-Phe-Asn-Pro-Leu-Ser-Arg-Lys-His-Gly-Gly-Pro-Lys-Asp-Glu-Glu-Arg-
horse	Ala-Gly-Gln-His-Phe-Asn-Pro-Leu-Ser-Lys-Lys-His-Gly-Gly-Pro-Lys-Asp-Glu-Glu-Arg-
yeast	Ala-Gly-Pro-His-Phe-Asn-Pro-Phe-Lys-Lys-Thr-His-Gly-Ala-Pro-Thr-Asp-Glu-Val-Arg-

	80 … 90 … 97
bovine	His-Val-Gly-Asp-Leu-Gly-Asn-Val-Thr-Ala-Asp-Lys-Asn-Gly-Val-Ala-Ile-Val-Asp-Ile-
human	His-Val-Gly-Asp-Leu-Gly-Asn-Val-Thr-Ala-Asp-Lys-Asp-Gly-Val-Ala-Asp-Val-Val-Ile-
horse	His-Val-Gly-Asp-Leu-Gly-Asn-Val-Thr-Ala-Asp-Glu-Asn-Gly-Lys-Ala-Asp-Val-Asp-Met-
yeast	His-Val-Gly-Asp-Met-Gly-Asn-Val-Lys-Thr-Asp-Glu-Asn-Gly-Val-Ala-Lys-Gly-Ser-Phe-

	100 … 100 … 117
bovine	Val-Asp-Pro-Leu-Ile-Ser-Leu-Ser-Gly-Glu-Tyr-Ser-Ile-Ile-Gly-Arg-Thr-Met-Val-Val-
human	Glu-Asp-Ser-Val-Ile-Ser-Leu-Ser-Gly-Asp-His-Cys-Ile-Ile-Gly-Arg-Thr-Leu-Val-Val-
horse	Lys-Asp-Ser-Val-Ile-Ser-Leu-Ser-Gly-Lys-His-Ser-Ile-Ile-Gly-Arg-Thr-Met-Val-Val-
yeast	Lys-Asp-Ser-Leu-Ile-Lys-Leu-Ile-Gly-Pro-Thr-Ser-Val-Val-Gly-Arg-Ser-Val-Val-Ile-

	120 … 130 … 137
bovine	His-Glu-Lys-Pro-Asp-Asp-Leu-Gly-Arg-Gly-Gly-Asn-Glu-Glu-Ser-Thr-Lys-Thr-Gly-Asn-
human	His-Glu-Lys-Ala-Asp-Asp-Leu-Gly-Lys-Gly-Gly-Asn-Glu-Glu-Ser-Thr-Lys-Thr-Gly-Asn-
horse	His-Glu-Lys-Gln-Asp-Asp-Leu-Gly-Lys-Gly-Gly-Asn-Glu-Glu-Ser-Thr-Lys-Thr-Gly-Asn-
yeast	His-Ala-Gly-Gln-Asp-Asp-Leu-Gly-Lys-Gly-Asp-Thr-Glu-Glu-Ser-Leu-Lys-Thr-Gly-Asn-

	140 … 150
bovine	Ala-Gly-Ser-Arg-Leu-Ala-Cys-Gly-Val-Ile-Gly-Ile-Ala-Lys
human	Ala-Gly-Ser-Arg-Leu-Ala-Cys-Gly-Val-Ile-Gly-Ile-Ala-Gln
horse	Ala-Gly-Ser-Arg-Leu-Ala-Cys-Gly-Val-Ile-Gly-Ile-Ala-Pro
yeast	Ala-Gly-Pro-Arg-Pro-Ala-Cys-Gly-Val-Ile-Gly-Leu-Thr-Asn

blue sensitized photooxidation) causes no inactivation.[148] This result is not unexpected. If the aromatic side chains were essential for catalytic function, enzymes from all species ought to contain them. Such is not the case. Horse dismutase has neither;[84] yeast enzyme has one tyrosine per subunit.[82,83]

D. Comparative Amino Acid Sequence Analysis

1. Introduction

The strong similarities in purification, stability, and chemical modification properties of Cu-Zn dismutases have an equally strong structural basis. The four complete amino acid sequences aligned in Figure 4[77-79,81,82,84] show a degree of homology comparable to that of the cytochromes *c*, indicating that enzymes of the Cu-Zn family are quite highly conserved and slowly evolving.[80] Alignment of identical residues (enclosed in boxes) requires introduction of gaps to compensate for the different numbers of residues in the subunit polypeptide chains (151 residues in the bovine protein and 153 residues in the others). Residues in one sequence, which are aligned against a gap in another sequence, suggest the presence of an additional small loop, in this region of the three-dimensional structure of the first protein. The amino terminus of the yeast dismutase is free, while that of all three mammalian dismutases is N-acetylated.

2. Conservation of Structural and Functional Determinants

The sequence comparisons in Figure 4 amply illustrate that evolving proteins exhibit the greatest homology in those segments containing residues essential for structure and function and exhibit greatest variability in sequence segments of lesser importance. For example, the single aspartic acid residue (Asp 81) and the six histidine residues (His 44, 46, 61, 69, 78, and 118) known to be ligands of copper and zinc in bovine dismutase[85,86] are strictly conserved, i.e., identical residues appear at homologous positions in the other three sequences. In several instances, the adjacent sequences are identical as well. Histidine residues which are not metal ligands in bovine dismutase are not conserved; in fact, yeast dismutase contains only histidine residues corresponding to the above six, and no others. The bovine sequence contains two nonactive site histidine residues, His 19 and His 41.[85,86] By analogy, it appears that the human sequence also contains two nonligand histidines, His 19 and His 110, while the horse sequence appears to contain four, His 15, His 19, His 43, and His 110. Other residues of known structural and functional importance in bovine dismutase are also identical, in homologous positions in the other sequences: Arg 141, whose modification by butanedione inactivates the enzyme,[144,145] Cys 55 and Cys 144, which form the intra-subunit disulfide bond,[79] and 19 of the 21 to 26 glycine residues in each subunit polypeptide, a number of which figure prominently in the sharp bends in the backbone.

While residues on the surface often show the greatest variation among proteins from different species,[119] such is not the case for the two loops, protruding from the back of the dismutase molecule, whose sequence conservation is high. Conservation in the

FIGURE 4. Complete amino acid sequences of copper-zinc SODs from (top to bottom) bovine erythrocytes, human erythrocytes, horse liver, and bakers' yeast. The numbers correspond to residue positions in the bovine sequence. The actual number of total residues in the subunit sequences is 151 for bovine and 153 for the other three dismutases. Half-cystine residues 55 and 144 in the bovine sequence, and homologous residues in the other three sequences, are in intrachain disulfide linkage. All other half-cystine residues exist as free cysteine. An amino terminal N_α-acetyl group is designated by "Ac". Residues which are identical in all four sequences are enclosed in boxes.[77-79,81,82,84] The primary structure of the human enzyme was analyzed independently in two laboratories.[80,81] The results differ at eight positions, six of which are amide assignments[80] (the sequence shown is taken from Reference 81). Independent determination of the yeast sequence in two laboratories[82,83] produced results which differ in two positions, both amide assignments[83] (the sequence shown is taken from Reference 82).

first loop (bovine residues 48 to 79) is not surprising, because it contains Cys 55, several metal ligands and residues 59 to 63, which lie on the rim of the superoxide-binding pocket. Conservation in the second loop (bovine residues 119 to 141) is less expected, although in its carboxyl terminal portion are Arg 141 and residues 134 to 139, which also form portions of the substrate-binding site.[134]

In conclusion, the nature of these homologies make it highly likely that these four dismutases, and presumably others in the Cu-Zn family, have very similar backbone conformations, featuring the β-barrel and the two loops of nonrepetitive structure observed in the bovine enzyme. Inferring three-dimensional similarity from strong sequence homologies has been shown to be valid for the cytochromes *c*, by solving the crystal structure of the cytochromes from horse and bonito.[149] For Cu-Zn dismutases, three-dimensional data are only available for the bovine enzyme, although crystallographic analysis has been begun with the Cu-Zn enzyme from spinach.[150] More detailed interspecies comparisons of the structure of the active site have been performed via spectroscopy (Section III.G).

3. Species Differences, Sequence Variations

It is gratifying to observe such sequence consistencies in catalytically and functionally important residues. However, sequence differences might be anticipated between these different species, because they do not all cope with oxygen toxicity in exactly the same fashion. In particular, yeast may grow anaerobically, while this option is not available to the three mammalian species. An examination of sequence variations may ultimately provide information about the structural basis of species differences in adaptation to an oxygen-containing environment.

The region from bovine residues 17 to 32 contains the longest stretch of sequence variability (Figure 4).[80] Expectedly, this sequence is close to neither the active site region nor the subunit interface, regions which merit high conservation. Interestingly, the variable segment is close in the three-dimensional structure to two other variable stretches, from bovine residues 94 to 98 and 107 to 111. All three segments are on the left and bottom side of the molecule (as shown in Figure 3A).[85,86] It has been proposed that this region may be an antigenic determinant of the molecule and that its "hypervariability" may be responsible for the low degree of immunological cross-reaction observed even between dismutases of closely related species[80,151] (see Section III.E). In addition to sequence differences, the conformation in the region corresponding to bovine residues 17 to 32 is expected to differ slightly in different species, because inserted residues will require additional looping or bends in the polypeptide. The bovine and yeast enzymes differ in their stability to urea,[98,151] and radical amino acid replacements in this variable region may in part be responsible.[83]

It is of interest to note that the single tyrosine or the single tryptophan per subunit, when present, is not always found in a homologous position. Thus the single tryptophan in human dismutase and the single tyrosine in yeast dismutase occur at residue 32 in each protein,[80-83] while the single tyrosine is at residue 108 in the bovine enzyme.[77-79] These two sequence locations are not immediately adjacent in the three-dimensional structure, although they both reside within the hypervariable locus, cited above. Considering the successes in hybridization of Cu-Zn dismutases (Section III.E), it should be possible to form a hybrid between a horse dismutase subunit, which lacks both tyrosine and tryptophan, and the subunit of a dismutase which contains one residue of tyrosine or tryptophan. If the horse dismutase subunit were specifically labeled with a fluorescent reporter molecule, then conformational changes between two distant sites on the dimer could be studied by fluorescence energy transfer measurements.[152]

One final observation about Figure 4 concerns the cysteine content. All four sequences have cysteine residues corresponding to the bovine intrachain disulfide. However, the yeast enzyme has no cysteine, while all three mammalian dismutases have a cysteine at residue 6. In bovine dismutase, Cys 6 is unreactive in the absence of denaturants[79] and is relatively inaccessible in the three-dimensional structure.[85,86] The human erythrocyte enzyme alone contains a second cysteine, at residue 111. As described earlier (Section II.C. 1.a), this cysteine residue is probably the progenitor of the presumed persulfide grouping, the 320 to 325-nm absorbance, and the lability of the human enzyme, compared to its bovine counterpart, when the McCord-Fridovich method is used for purification.[30,31]

E. Hybridization and Immunological Cross-Reactivity

Structural relatedness can be evaluated by hybridization and immunological studies. A principal attraction of these methods is that they are considerably less tedious than sequence determination and thus are suited for surveying dismutases from a number of species. In addition, they measure parameters of "relatedness" different from those reflected in sequence comparisons. Hybrids are Cu-Zn dimers in which the two subunits are from different species. Hybrid formation consequently reflects a property of the interior of the molecule: the evolutionary conservation of the subunit interface and the inter-subunit binding forces. By contrast, immunological cross-reaction reflects the molecular exterior, because antibodies are raised to surface features of the enzyme, when the native protein is used as an antigen.

The criterion for hybridization has been the appearance of a new electrophoretic band, with mobility intermediate between that of the two parental species, following exposure to mild denaturing conditions. Brief heating,[153] or treatment with carbon tetrachloride[154] or urea[155,156] have all been successfully used to promote subunit interchange.

Hybridization, it turns out, is scarcely a gamble. It appears to have been successful whenever attempted, regardless of phylogenetic distance between the two species involved. Hybrids are formed between species as distant as human and swordfish,[153] guinea pig and salmon, *Neurospora* and guinea pig,[157] and wheat germ (isozyme I) and bovine,[155] or between species as close as human and mouse[154] or human and guinea pig.[153] The unavoidable conclusion is that the subunit interface is highly conserved among Cu-Zn dismutases.

Retention of catalytically important residues is certainly the most visible manifestation of conservative trends in protein evolution. However, certain structural traits of the protein molecule may be just as essential for biological function, and consequently just as strongly conserved. The dimeric structure of Cu-Zn dismutases is such a trait (Table 1), and its invariance and the apparent invariance (judged by hybridizations) of the forces which maintain it, is not too surprising. Similar evidence has been obtained for compatibility of subunits of glyceraldehyde-3-phosphate dehydrogenase (a tetramer), from species as diverse as *E. coli* and pig.[158]

Immunological cross-reactivity has been examined by several criteria: inhibition of dismutase activity, Ouchterlony double diffusion precipitin lines, and radioimmunoassay. By these methods, the relatedness of dismutase structure is much less than judged by hybridization. Dismutases from closely related species do show cross-reactions, e.g., spinach antibody with extracts from seed plants, mosses, and ferns,[159] and rat antibody with purified human, bovine, or horse dismutases.[153,160] (There is a report that anti-rat liver dismutase antibody does not cross-react with bovine erythrocyte SOD.[36]) However, more distant relatives do not cross-react, e.g., spinach antibody does not cross-react with bovine dismutase,[159] nor bovine antibody with swordfish dismutase.[160] A rather remarkable specificity is shown by antibody to the bacteriocuprein from *P. leiognathi*, which neither cross-reacts with purified Cu-Zn dismutases from fish (e.g., marlin, trout, red snapper, sea bass) nor with the dismutase from its own host pony fish.[11] One may conclude that while certain surface regions of Cu-Zn dismutases (e.g., the catalytic site) are highly conserved, other regions vary in charge, hydrophobicity, or conformational nuances, thus generating antigenic differences. Amino acid sequence variability suggests that some of these species-specific antigenic determinants are in the regions corresponding to bovine residues 17 to 32, 94 to 98, and/or 107 to 111, as noted earlier (Section III. D.3).

F. General Features of the Catalytic Mechanism

1. Copper Reduction and Reoxidation

The very nature of the dismutation of superoxide to dioxygen and hydrogen peroxide creates a complication for all would-be physiological catalysts. At neutral pH superoxide radical is predominantly in the anionic form, $O_2^{\cdot-}$, being the conjugate base of the hydroperoxyl radical ($HO_2\cdot$), whose pK_a is 4.8.[161] Any catalyst must circumvent the electrostatic repulsion and consequent slow reaction between two superoxide anions indicated in the chemical equation (Equation 2).[5]

$$O_2^{\cdot-} + O_2^{\cdot-} + 2H^+ \longrightarrow H_2O_2 + O_2 \qquad (2)$$

The apparent ploy used in catalysis by free cupric ion is to eliminate the negative charge by forming a metal-superoxide product, e.g., $Cu\text{-}O_2^+$. The enzymatic ploy is different: copper acts as an electron carrier between the superoxide radicals, in two successive half-reactions, and the two anions never actually meet. The enzyme mechanism is illustrated in Equations 3 and 4.[5,6,8,76]

$$E\text{-}Cu^{2+} + O_2^{\cdot-} \longrightarrow E\text{-}Cu^{1+} + O_2 \qquad (3)$$

$$E\text{-}Cu^{1+} + O_2^{\cdot-} + 2H^+ \longrightarrow E\text{-}Cu^{2+} + H_2O_2 \qquad (4)$$

(E = one dismutase subunit)

The resting enzyme contains copper in the oxidized, Cu^{2+}, state, deduced from X-ray absorption studies.[141] In the first half-reaction, enzyme-bound copper is reduced from the cupric to the cuprous state, and the first superoxide radical is oxidized to dioxygen (Equation 3). In the second, copper is reoxidized to cupric, while the second superoxide is simultaneously reduced and protonated to form hydrogen peroxide (Equation 4). Thus the first half-reaction involves only electron transfer, while the second requires both electron and proton transfer. The source of the proton is not known for certain, and has been the subject of some investigation (see Section III.G.5).

The principal evidence for this cyclic mechanism, of alternate reduction and reoxidation of copper, is from kinetic studies using pulse radiolysis. If an oxygenated aqueous solution, containing formate, is irradiated with a brief electron pulse (about 1.5 μsec duration), superoxide anion will be the only radical present, within about 1 μsec after irradiation. The initial $O_2^{\cdot-}$ concentrations, thus generated by pulse radiolysis, are approximately 1 to 20 μ*M*.[46,96] From monitoring disappearance of $O_2^{\cdot-}$, by UV absorbance measurements of irradiated enzyme solutions, the second-order rate constant for catalysis is found to be about 2×10^9 M^{-1} sec^{-1}. An identical value has been found for both bovine[96,162,163] and human[164] erythrocyte dismutases, and the constancy of this value has been shown over the range of pH 4.8 to 9.5 for the bovine enzyme[96,162] and pH 5.7 to 10.5 for the human enzyme.[164] The enormity of this rate constant, plus studies of its dependence upon temperature[163] and viscosity,[162] strongly suggest that the rate of the enzyme-catalyzed reaction is diffusion limited. From monitoring at a wavelength near the absorbance maximum of the enzyme-bound copper, it is possible to measure (1) the rate at which $O_2^{\overline{\cdot}}$ reduces the oxidized (i.e., "resting") enzyme to the cuprous form, and (2) the rate at which $O_2^{\overline{\cdot}}$ oxidizes the reduced enzyme (formed by treating resting enzyme with hydrogen peroxide) to the cupric state. The equality of these two rate constants, and their identity with that for superoxide dismutation, indicates that alternate reduction and reoxidation are concomitant with catalysis.[163]

X-ray absorption spectroscopy has definitively established that the valence of Zn^{2+} does not change during catalysis, and that on reduction, the valence of enzyme-bound copper changes from Cu^{2+} to Cu^{1+}.[141] If superoxide anion is generated by reduction of dioxygen at a dropping mercury electrode, concentrations of the radical much higher than those produced by pulse radiolysis can be generated. With a polarographic assay based upon this principle, a K_m value of 0.36 m*M* is determined for the substrate superoxide, with the bovine erythrocyte dismutase.[8,165]

Some evidence has been presented in opposition to the proposed mechanism, that enzyme-bound copper undergoes alternate reduction and reoxidation. Aqueous solutions of bovine dismutase were irradiated with γ-rays at 77 K to generate the superoxide radical *in situ.* Both superoxide and the cupric form of dismutase are paramagnetic. Electron paramagnetic resonance (EPR) spectra revealed a new signal, attributed to a "loose complex" of the hydroperoxyl radical ($HO_2\cdot$) with enzyme copper.[166] In spectra recorded at successively higher temperatures, the signal of this $HO_2\cdot$ species decayed before the oxidation state of copper changed.[167] That is, dismutation of superoxide preceded reduction of cupric enzyme, or oxidation of cuprous enzyme, depending upon which was initially present. The interpretation of these data was that dismutation might proceed catalytically without a change in the oxidation state of copper.[167] Quite simply, this interpretation is anathema to the existing dogma about the catalytic mechanism of Cu-Zn dismutases and is difficult to reconcile with pulse radiolysis experiments, which determine an identical rate constant for dismutation and reduction (or reoxidation) of the enzyme.[163] The proposed assignment of the new EPR signal to a "loose complex" of $HO_2\cdot$ with enzyme copper deserves further scrutiny. The proximity of $HO_2\cdot$ and Cu^{2+} in such a complex would be conducive to spin coupling and consequent disappearance of their EPR signals. The persistence of both Cu^{2+} and $HO_2\cdot$ EPR signals is explained by the "loose" nature of the complex.[166] However, there is an alternative explanation: that the new signal is not from a catalytically important species, but from superoxide radical which is distant from the enzyme active site, and that the observed decay of this signal, which precedes valence changes in copper, is a nonenzymatic process which is not directly relevant to the enzymatic mechanism.

2. Inner Sphere Electron Transfer

In principle, electron transfer between superoxide and copper might occur with the superoxide in either the inner or the outer coordination sphere of enzyme-bound copper. The available evidence (obtained with the bovine dismutase and reviewed in References 8 and 168) supports a mechanism in which superoxide anion displaces a water molecule from the inner coordination sphere of copper, prior to or during electron transfer, in both the reduction and oxidation steps of dismutation. The evidence is: (1.) Cyanide is an anionic inhibitor which is known from spectroscopic studies to enter the inner coordination sphere of enzyme-bound copper,[169] and known from kinetic studies (using the polarographic method for assay)[165] to be a competitive inhibitor. Hydroxide ion is also a competitive inhibitor of the enzyme.[8,165] (2.) Studies of the effect of oxidized protein on the relaxation of water protons demonstrate a rapidly exchanging water molecule in the first coordination sphere of Cu^{2+}.[170] (3.) The presence of a corresponding open coordination position in the reduced protein is indicated by studies of the influence of cuprous dismutase on the relaxation of the ^{35}Cl nucleus of chloride ions.[171]

3. The Question of Half-of-the-Sites Reactivity

A reasonable expectation from the mechanism of alternate reduction and reoxidation is that equal concentrations of the cupric and cuprous forms of the enzyme will be present during steady-state turnover of superoxide. In one of the same studies which laid the foundation for this mechanism, evidence was presented that this expectation was not met, and that the ratio of the two forms was significantly different from 1:1, regardless of whether the steady state was approached from the reduced or the oxidized protein.[163] As explanation, a mutually inhibitory interaction between the copper sites on the two subunits of the dismutase dimer (a half-of-the-sites reactivity) was proposed, even though the considerable distance between the two sites (some 34 Å in the crystal structure)[85,86] argued against any interaction between them.[163]

Several years later, this explanation was withdrawn; the inequality of oxidized and reduced states was attributed to contamination of the enzyme with an inactive form of the protein, with EPR properties similar to the active form. In the same publication, a much closer agreement with the expected 1:1 ratio was reported from experiments using potassium superoxide, rather than pulse radiolysis, as a source of the superoxide radical.[172] The current status of this issue is that the two active sites of bovine dismutase function independently and simultaneously without any mutually inhibitory interactions during catalysis.[172]

Although the hypothesis of half-of-the-sites reactivity has now been laid peacefully to rest, two interesting studies, reflecting its halcyon days, merit mention. One study used a chemical approach to address the question of subunit interaction. (The answer was negative.)[156] Bovine dismutase was inactivated by successive modifications with diazonium 1H-tetrazole and hydrogen peroxide, then incubated in urea with unmodified enzyme. From the reconstituted and renatured mixture, a hybrid enzyme was isolated, containing one subunit each, from modified and unmodified proteins. The hybrid had a specific activity equal to one half that of unmodified protein, suggesting that the two subunits in native enzyme act independently and contribute equally to the catalytic turnover.[156]

The second study is of interest, however, because it does provide evidence for an interaction between subunits, which occurs during reconstitution with copper.[173-175] Zinc-containing, copper-free bovine dismutase was incubated with increasing amounts of Cu^{2+}, and the relative amounts of copper-free, mono- and di-copper dismutases were determined, following their separation by acrylamide gel electrophoresis.[173] (This behavior is to be contrasted with that of Mn/Fe dismutases where apo- and metal-containing forms appear to have the same electrophoretic mobility.)[102,103] Equilibrium[173,174] and kinetic studies,[175] of both spectroscopic and catalytic parameters, indicated (1) positive cooperativity in copper binding, ie., the binding of Cu^{2+} to the first subunit facilitated binding to the second subunit of the dimer by an estimated 0.4 to 0.6 kcal/mol[173] and (2) negative cooperativity in regain of activity, i.e., activity per copper site was smaller in the di-copper than in the mono-copper protein.[174] These observations are suggestive of interactions between the copper-binding sites on the two subunits of the bovine dismutase dimer, a distance of some 34 Å.[85,86] The propagation of protein conformational changes over such large distances has been demonstrated in other systems, e.g., in certain receptor proteins.[176] While conformational communication between the two active sites may well occur during reconstitution, this is no assurance of active site interaction during catalysis by the metal-replete enzyme. Furthermore, with a rate constant for the enzyme-catalyzed reaction, which is diffusion limited,[162,163] conformation changes over a large portion of the molecule might not be sufficiently rapid to be catalytically significant.

4. The Search for the Elusive Active Subunit

There are a number of publications presenting evidence for catalytic activity in isolated, dissociated subunits of Cu-Zn SOD. These data have been evaluated and found to be inconclusive; there is no definitive evidence that Cu-Zn dismutases retain catalytic activity under solution conditions which dissociate the polypeptide chains of the dimer.[95,151] Nonetheless, this section is dedicated to just these data. Though they are equivocal, they are penetrating pedagogical examples of some of the limitations in measuring activity in denaturing media and in deducing molecular weight in denaturing solvents by empirical approaches. (1) Several Cu-Zn dismutases have been shown to retain full activity in 8 *M* urea,[98,151,177] the bovine enzyme, for a period of days.[95] The molecular weight of these dismutases in 8*M* urea has been determined to be about 16,000, i.e., the molecular weight of the isolated subunit, using molecular sieve chromatography in this solvent as an empirical method of molecular weight determination.[151,177] The objection that 8 *M* urea might not actually dissociate the dimer was confronted by additional evidence: that hybridization, between wheat germ isozyme II and either bovine or yeast Cu-Zn dismutase, was accelerated in 8 *M* urea.[155] These data have several pitfalls.[95,151] First, separations on molecular sieve columns are based upon differences in molecular size, or effective molecular radius, and not molecular weight. The elution volumes from molecular sieve chromatography will be accurate measures of molecular weight only if the hydrodynamic character of the protein being analyzed is congruent to that of the standard proteins used to calibrate the column. In 8 *M* urea, most Cu-Zn dismutases evidently remain compact molecules, compared to the flexible, loose structure of the denatured calibration proteins. Dismutases thus elute at later volumes, corresponding to erroneously low apparent molecular weight. Fortuitously, this low value is very close to the subunit molecular weight. Sedimentation equilibrium in the ultracentrifuge is a more rigorous method for molecular weight determination. By this method, bovine dismutase in 8 *M* urea has the native molecular weight of about 32,000.[95] Furthermore, a facilitated formation of hybrids is not evidence of dissociation of subunits. Hybrids may be formed, even if the extent of dissociation is very small, as long as the stability of the hybrid is comparable to that of the parental dimers, and there exists a rapid equilibrium between dimers and subunits.[95,151] This principle, describing the effect of dissociation equilibrium on hybridization was reported approximately 20 years ago in studies of hybridization among hemoglobins.[178,179] (2) Wheat germ isozyme I is dissociated by sodium dodecyl sulfate.[19] The monomeric band obtained after acrylamide gel electrophoresis in detergent was shown to have activity by the enzyme staining technique.[180] If sodium dodecyl sulfate is removed from isozyme I, the activity is slowly regained. This activity was attributed to the monomeric species, because samples of partially reactivated enzyme, analyzed by detergent gel electrophoresis, are almost entirely monomeric.[181] The enzyme-staining technique involves soaking the gel,[182] in a presumably detergent-free solution, and therein lies the rub in the experiment. During the soak, there may have been sufficient dilution of sodium dodecyl sulfate to allow reassociation of subunits to the dimeric form, which may well be the source of the observed activity.[180] In the electrophoretic analysis of the partially reactivated enzyme,[181] the problem confronted is the application of a transport or separation stress to a system in association-dissociation equilibrium. If the electrophoretic separation of monomers from dimers is sufficiently fast, then the monomer band might be significant, even though active dimer were a major species in solution prior to electrophoresis.

G. Fine-Structure Analysis of the Metal-Binding Sites by Spectroscopy

1. Introduction

Spectroscopic methods began to provide information about the metal sites of Cu-Zn dismutases well before the crystal structure was solved and continue to be vehicles of increasingly detailed descriptions, well after the 3Å structure was reported. Their success has relied in large part upon the preparation of enzyme derivatives in which the native metal(s) has been removed, or has been replaced by another metal, with more convenient spectroscopic properties. This section will begin with a brief discussion of the principles of metal reconstitution and of the spectroscopic techniques themselves. The main charge of the section will be the review of three active areas of spectroscopic analysis of the structure and mechanism of the active site: the structural role of zinc, the presence and function of the bridging imidazolate, and the assignment of proton nuclear magnetic resonances to specific imidazole nitrogen or carbon atoms. Spectroscopic studies have almost exclusively examined the Cu-Zn dismutase from bovine erythrocytes, but recently have begun to be extended to enzymes from other species.

2. Principles of Metal Replacement

Like chemical modification, metal replacement studies seek to introduce a specific change in a protein functional group. In both approaches, demonstrating that the change occurs at a specific site, without significant structural alterations elsewhere in the protein, is an absolute necessity. Experimentally, this requirement translates into a concern about (1) the method of forming apoprotein and (2) the method of reconstitution (because most of the metal substitution derivatives have been formed by this two-step process, via the apoprotein).

The apoprotein is most frequently prepared by incubating the native enzyme at acid pH, usually in the presence of a chelating agent.[1,87,183,184] The bovine protein does bind EDTA, which must be removed by dialysis vs. high concentrations of sodium perchlorate[184] or other salts[140a,185] to prevent spurious metal binding in subsequent reconstitution. The pH at which the metal ion is added back to the apoprotein is crucial, and the rate of addition of metal ion may also be important.[87,183] Typically, specific replacement of the zinc ion has been accomplished by adding cupric ion to the apoprotein at acid pH, then dialyzing the product vs. pH 7.8 buffer, and adding the divalent ion substitute of zinc.[87,183] This protocol owes its success to the fact that only the copper-binding site is intact at acid pH. When the pH is raised, the zinc-binding site then forms and becomes competent for liganding zinc or other divalent cations.[186]

Using the proper conditions, it is possible to reconstitute apoprotein with copper and zinc and to obtain a fully active protein, with spectroscopic properties identical to those of native enzyme.[87,183] This important "metal replacement control" demonstrates that no irreversible structural changes are introduced by the reconstitution conditions. Metal replacements have been obtained for either zinc, copper, or both. Copper is essential for activity; all metal replacements of it have been totally inactive.[1] Zinc, on the other hand, is dispensable. Thus copper-containing proteins, in which zinc has been replaced by divalent mercury, cobalt, or cadmium, have activities identical or nearly identical (70 to 90%) to that of native copper-zinc protein.[183] The activity of these zinc replacement derivatives (1) is consistent with other evidence that the catalytic mechanism depends principally upon the copper,[141,163] and (2) gives further credence to the suitability of the depletion-reconstitution, two-step process for replacing metals without perturbation of the protein structure. In fact, a mercury-for-zinc substitution derivative was used as a heavy atom derivative in solving the crystal structure.[85]

Quantitative metal analyses are crucial, not only for establishing the stoichiometry of metal binding after reconstitution, but also in defining the extent of metal depletion in the apo form. Obtaining evidence for the specificity and the site of metal replacement is equally important. Metal titration data or integral stoichiometries (two metal ions per molecule of Cu-Zn dimer, or one per polypeptide subunit) and high specific activites (for copper-containing derivatives in which only zinc has been replaced) are convincing, albeit circumstantial, evidence for the absence of nonspecific binding and for specific substitution at the zinc site, respectively.[87,183,184] Probably more convincing evidence for substitution at the zinc site is the demonstration of competition: that the replacement cation will not bind at all, if apoprotein has previously been reconstituted with zinc alone.[184]

3. Spectroscopic Methods: Some Principles and Early Applications

The principal spectroscopic methods which have been applied to Cu-Zn dismutases are electron paramagnetic resonance (EPR) spectroscopy, nuclear magnetic resonance (NMR) spectroscopy, and to a lesser extent, absorption spectroscopy. Spectroscopic techniques for studying metalloproteins have been classed as either "metal-centered" or "ligand-centered".[187] depending upon whether electromagnetic absorption in the metal or in the protein, respectively, is responsible for the spectrum. Absorption spectroscopy involves transitions between electronic energy levels.[188] Used in the visible region of the spectrum, absorption spectroscopy is a "metal-centered" technique; copper and Co^{2+} have absorption bands in this region.

Both EPR and NMR spectroscopy reflect resonance absorption by elementary particles in a magnetic field. EPR has been a metal-centered technique, because EPR spectra reflect transitions between energy levels of unpaired electrons.[188] Among the metals commonly studied in dismutases, only Cu^{2+} (but not Cu^{1+}) and Co^{2+} show EPR signals. The requirement for energy absorption in NMR spectroscopy, a nucleus with nonzero spin angular momentum,[188] is met by hydrogen (proton), ^{1}H, deuterium, ^{2}H, and the cadmium isotope,^{113} Cd^{2+}. Applications of NMR spectroscopy therefore have been both "metal-centered" and "ligand-centered".

The sensitivity of EPR and NMR spectra to the chemical environment of the absorbing species has made these two techniques particularly useful probes. (1) Features of the EPR spectrum of bovine dismutase suggested that the cupric copper was liganded by nitrogen functions[169,183,189-191] before chemical modification and crystallographic data proved four histidine imidazole nitrogen functions to be the ligands. Similar first-derivative EPR spectra were found for Cu-Zn dismutases from cow, chicken, yeast, and *Neurospora*. Differences in the second derivative superhyperfine patterns suggested that the fine structure of the metal-binding sites of bovine and chicken enzymes were similar to one another, but different from those of the fungal enzymes from yeast and *Neurospora*.[183] (2) If a molecule contains two paramagnetic species close together, their unpaired spins may couple by a magnetic dipole interaction, resulting in loss of the EPR signal at low temperature. This magnetic coupling principal has been used as evidence for the proximity of the copper and zinc sites in bovine dismutase, before crystallography revealed their 6 Å separation:[168] the characteristic EPR spectra were absent in protein reconstituted with (a) Cu^{2+} in both the copper and the zinc sites,[189,192] and (b) with Cu^{2+} in the native copper site and Co^{2+} in the native zinc site.[184,193,194] (3) Many of the more recent EPR and NMR applications have compared spectra of native enzyme, apoprotein, and apoprotein reconstituted in either the zinc site alone or in both sites. These applications have been used for studying the role of zinc, the bridging imidazolate, and in assigning NMR resonances to protons of specific residues. In either "metal-centered" or "ligand-centered" NMR studies, copper must be present in the cuprous form (i.e., "reduced protein"). The paramagnetism of cupric copper is an efficient relaxation source, and signals from protons within about 13 Å are broadened to the point where they blend into the background noise.[185,195,196]

4. The Role of Zinc in Stabilizing the Active Site

It has been apparent since the discovery of SOD that zinc does not directly participate in the catalytic process. Apoprotein reconstituted with copper alone possessed substantial activity.[1] While the blue color of the enzyme implicated copper, assays for zinc were not performed until after bovine dismutase was discovered.[197] The zinc ion was, however, ascribed a role in maintaining the stability of the protein. The evidence presented was (1) apoprotein, reconstituted with copper alone, was less thermostabile than protein reconstituted with both copper and zinc,[94] while (2) copper-free protein, containing zinc as the only metal, was more resistant to dissociation by sodium dodecyl sulfate than the apoprotein.[198] Additional information about the structural role of zinc has come from spectroscopic techniques.

The proton NMR spectrum of most proteins may be divided into an "upfield" (higher resonance energy) region, containing resonance peaks of protons bonded to aliphatic functional groups, and a "downfield" (lower resonance energy) region of resonances from protons bonded to aromatic functionalities.[199] Within the aromatic envelope are peaks from imidazole N-H protons and imidazole C-2 and C-4 ring protons.[185,187,195] (See the structure below for numbering of the imidazole ring.)

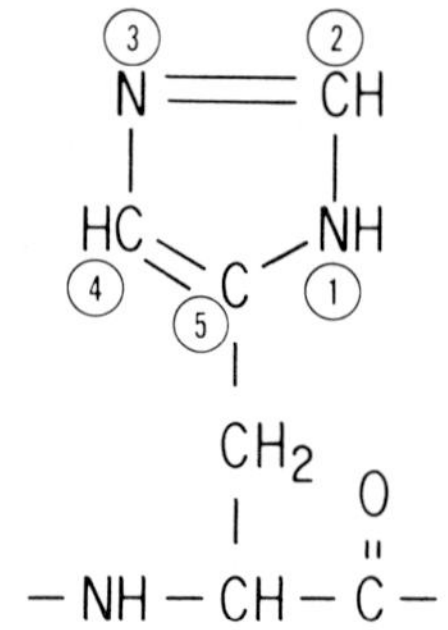

The N-H protons are exchangeable. No resonance peaks will be observed for N-H protons when there is rapid exchange of the proton with water, as there is, for example, in the free amino acid, histidine, or in a histidine residue which is totally exposed to solvent. If exchange is prevented or significantly reduced in rate, for example, by inaccessibility of the imidazole side chain or by hydrogen bonding or metal liganding involving the N-H group, then a resonance peak will be present.[185,200]

The reduced (cuprous) holoenzyme of bovine erythrocyte dismutase shows several discrete peaks in the imidazole N-H region of the spectrum, corresponding to N-H functions which ligand to copper or zinc. These resonances are not seen in the spectrum of apoprotein, because the N-H functions are exposed to solvent and freely exchangeable in the absence of metals. If apoprotein is reconstituted with zinc alone, at one equivalent per subunit, the N-H resonances reappear. In fact, in this region the spectrum of the zinc-only protein closely resembles that of native (reduced) enzyme. These data have been presented as evidence of a "zinc-promoted organization of the active site structure", i.e., that zinc alone can stabilize the histidine ligands in a coordination structure very similar to that found in the active site of holoenzyme, which contains both copper and zinc.[185] In this same study, chemical modification evidence corroborated that the accessibility of histidine residues is dependent upon the presence of zinc. If apoprotein is reconstituted with increasing amounts of zinc (i.e., zero to one equivalents per subunit), the N-H resonances progressively reappear, and their reappearance correlates well with a decrease in the amount of histidine which can be modified by diethylpyrocarbonate. From a maximum modification of eight (of the eight) histidine residues in each subunit (in the apoprotein), diethyl pyrocarbonate reaction decreases to zero residues reactive (in the zinc-replete protein).[185]

This approach has been extended to a comparison of bovine Cu-Zn dismutase with wheat germ dismutase isozymes I and II.[200] Based principally upon the NMR spectrum in the region of exchangeable imidazole N-H protons, all three Cu-Zn dismutases appear to have highly similar active site structures, with only slight differences in the microenvironment of histidine ligands, evinced in slight shifts in the resonance positions of homologous peaks. In all three, reconstitution with zinc alone (one equivalent per subunit) transforms the flexible and solvent-accessible active site region of the apoprotein into a structure highly resembling that of the holoenzymes. These data greatly enhance the argument for the evolutionary conservation of active site structure, deduced from amino acid sequence comparisons (Section III.D.2). Wheat germ isozyme I is markedly more sensitive to dissociation by sodium dodecyl sulfate than the other two dismutases;[19,180,181] clearly, the structural explanation for this stability difference must be found outside of the metal-binding sites of these proteins, which are shown by these NMR studies to be highly homologous.

Similar conclusions about the organizational role of zinc have been drawn from proton NMR studies of the aliphatic regions of proton NMR spectra. Reconstitution with zinc alone changes the spectrum of bovine apoprotein into one similar to that of holoenzyme.[139] Given that CD spectra suggest that the β-sheet structure is retained in apoprotein,[137] it may be that the zinc provides a nucleus around which the preformed secondary structure of the metal-free protein subunit can fold, forming an active site conformation.[139]

Zinc ion definitely can organize the active site structure within the apoprotein, but does it really serve that role in the holoenzyme? Assaying the activity of enzyme containing copper alone could give an answer. If copper-only enzyme had 100% activity, then the structural importance of zinc might be viewed askance. Answering the lead question has been befuddled by the variable amounts of activity which have been measured for different preparations of copper-only protein.[183] Recent EPR and absorption spectra have provided some insight into this activity variation, and have implied that the structural importance of zinc may not be as firmly established as thought previously.[201]

The spectroscopic data indicate that in apoprotein, reconstituted only with copper (two equivalents per dimer), the copper can migrate from a copper site in one subunit, to the empty zinc site in another subunit. This migration will form a metal-free subunit and a subunit with copper in both the native copper and the native zinc-binding sites. The rate of this proposed migration increases with increasing pH. The authors propose that apoprotein, reconstituted with copper alone and containing copper exclusively in the native copper sites, has 100% of the holoenzyme activity.[201] Any observed activity values less than 100% are attributed to migration and formation of metal-free (inactive) subunits, during the exposure to neutral-to-alkaline pH solutions following the reconstitution treatment. This migration phenomenon also decreases the strength of arguments for a structural role of zinc based on decreased thermal stability of the copper-only protein.[94] The decreased stability could likewise be attributed to formation of metal-free subunits or to apoprotein (which is less thermostable than holoenzyme) rather than to a stabilizing function for zinc.

In summary, we find a conundrum in resolving the question about the functional importance of zinc. Addition of zinc to the apoprotein clearly facilitates formation and stabilization of the active site. However, addition of copper alone to apoprotein is sufficient for appearance of total dismutase activity. Dismutases of the manganese/iron family have survived quite satisfactorily with just one metal in their active site. Why has the Cu-Zn dismutase, which evolved at a later time, acquired two metals in its catalytic center, one of which is clearly not required for dismutation? Perhaps a combination of evolutionary comparisons and spectroscopic analysis will provide insights into this matter.

5. Presence and Function of the Bridging Imidazolate

The crystallographers[85,86] confirmed the spectroscopists' prediction that the copper and zinc ion in each subunit of bovine dismutase were in close proximity.[184,189,192-194] The 6Å space between the two metals is spanned in the crystalline enzyme by the imidazole of His-61, although the ring is not co-planar with the metals.[86,87] Shortly after the crystal structure was unveiled, a functional role was hypothesized for this bridging imidazolate.[142] A certain mystique had always surrounded the second step of the dismutase reaction, in which superoxide is reduced to hydrogen peroxide. This step requires not only transfer of electrons, but also transfer of protons. The lack of a D_2O effect on activity (in the xanthine oxidase-cytochrome *c* assay) suggested that bulk water was not the source of these protons.[142] The invariance of the second-order rate constant for the enzyme-catalyzed reaction, over the range from pH 4.8 to 9.5,[96,162,164] eliminated most amino acid side chains as potential proton donors.[142] The hypothesis about the bridging imidazolate proposed that the bridge was broken when Cu^{2+} was reduced to Cu^{1+}, and that the N-H group thus formed was the proton source in the O_2 reduction step (see Equation 5).

(5)

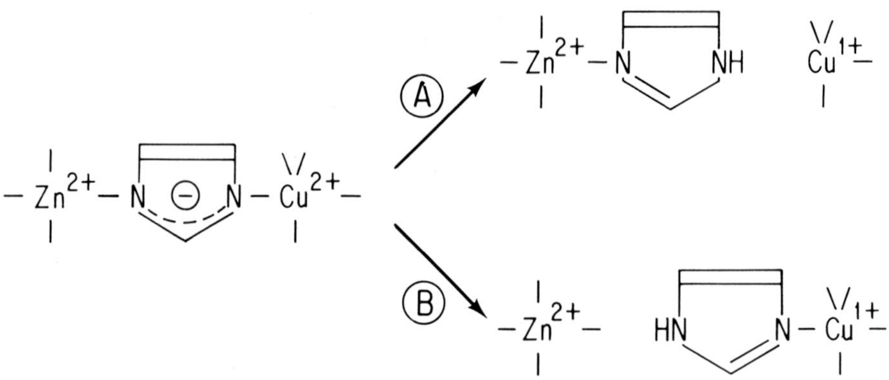

The spectroscopic experiments described below have evaluated this hypothesis and have determined whether the copper-facing or the zinc-facing imidazole nitrogen is broken on reduction of cupric copper (Pathway A or Pathway B in Equation 5).

The principal model for these studies has been the Co^{2+}-Cu^{2+} derivative of bovine dismutase. The metal stoichiometry, the prevention of cobalt binding by prior reconstitution with zinc, and the high specific activity strongly suggested that this analog had the intended constitution: cobalt in the native zinc site and copper in the native copper site.[183,184] (Note, however, that cobalt appears to bind at the native copper site, if zinc-only protein is incubated with cobalt at alkaline pH.)[202] Establishing the identity of the Co-Cu derivative is crucial, because its properties are relevant to those of holoenzyme only if the substituted cobalt is strictly analogous to the zinc ion in the native enzyme.

The proximity of zinc and copper ions was concluded from the magnetic coupling between paramagnetic Co^{2+} and Cu^{2+} in this dismutase derivative. Cobalt-only protein showed an EPR spectrum typical of tetrahedral Co^{2+}; this signal was very broad and weak in the Co^{2+}-Cu^{2+} protein, but reappeared in the Co^{2+}-Cu^{1+} protein (obtained by reduction with dithionite).[168,184,193] While this spin coupling between Co^{2+} and Cu^{2+} did establish proximity, it did not establish that cobalt and copper, and by inference zinc and copper, were both actually liganded to the bridging imidazole of His 61. Magnetic coupling may be mediated by a through-space mechanism (i.e., in the absence of a bridging ligand) or by a through-bond mechanism (i.e., with ligand bridging, for example by overlap of orbitals from each metal with the aromatic system of the imidazole). This uncertainty was resolved by magnetic susceptibility measurements.[168,203,204] The spin-pairing energy, of the Co^{2+} and Cu^{2+} spins is an important factor in this technique. The pronounced deviation of the magnetic susceptibility data from the behavior predicted for distant metal sites indicated antiferromagnetic coupling between the two metals. At a distance of some 6 Å, a through-space mechanism for this type of coupling is unconscionable, and a through-bond mechanism is strongly indicated, thus implicating a bridging imidazolate, to which both metals ligand. Recently, further evidence for a copper-zinc imidazolate bridge has been reported from electron spin echo spectroscopy of the bovine erythrocyte holoenzyme, a technique which reflects the interaction of the Cu^{2+} spin and the nonliganded ^{14}N atom of a ligated imidazole.[204a] The bridged metal pair is present at neutral and alkaline pH, and absent at pH 3.1.

Interestingly, the optical and EPR spectra of Co^{2+}-only protein (i.e., no copper present and therefore no copper liganding to His 61) were very similar to those of reduced (Co^{2+}-Cu^{1+}) protein.[8,168,184,193] This similarity suggests that in the reduced Co protein, His 61 is not liganded to copper, but only ligands to Co^{2+}, and by analogy, only to zinc in the reduced native enzyme. Put another way, these data suggest that the copper-facing imidazolate bond is broken prior to or during the reduction step of dismutation. This suggestion has been confirmed by NMR studies of the cadmium-113-copper enzyme (copper in the native copper site and ^{113}Cd in the native zinc site).[205] When the ^{113}Cd NMR signals of cadmium-only protein and reduced copper protein ($^{113}Cd^{2+}$-Cu^{1+}) are compared, the line width of the two are the same, and only small differences are observed in the chemical shift of the signals. These observations demonstrate that the bridging imidazolate is in the same state in the Cd-only and in the reduced cadmium-copper protein, i.e., that the cadmium- (and by inference zinc-) imidazolate bond remains intact on reduction, while the copper-facing bond is broken. In conclusion, the data uniformly support Pathway A in Equation 5. Pathway A, much more than Pathway B, should facilitate the expected change in ligand geometry, from pentavalent for cupric copper to tetrahedral for cuprous copper.[8]

At this juncture it is important to inquire whether breaking the copper-facing His 61 bond is actually relevant to the catalytic mechanism. This question has been answered in the affirmative by pulse radiolysis studies of Co-Cu bovine enzyme.[206] Monitoring at a visible wavelength characteristic of Co^{2+}, the second-order rate constant for the absorbance change is found to be identical to that for dismutation of superoxide. Thus spectroscopic changes in the environment of the cobalt center, presumably reflecting breaking or reforming of the imidazolate bond with copper, are concomitant with catalysis, i.e., these changes occur at a rate sufficiently rapid to be catalytically significant and part of the catalytic mechanism. (The catalytic rate constant, for dismutation of superoxide radical, is the same in cobalt-copper and in holoenzyme, suggesting that analogies between this metal derivative and native enzyme are valid.)[206]

One study of the Co-Cu derivative of bovine dismutase reported that at acid pH, the cobalt-facing imidazolate bond was broken (and by inference that the zinc-facing imidazolate bond would be broken at acid pH in the native enzyme).[207] A clearer understanding of this phenomenon has come from subsequent spectroscopic studies of the pH dependence of metal binding.[186,208] Zinc and other divalent cations in the native zinc binding site are labile at acid pH (below about pH 4.5). To be sure, the zinc-facing imidazolate is broken at low pH, but the mechanism is not simple protonation of the imidazole nitrogen (shown in Equation 5, Pathway B). Rather, the bond is broken because the zinc-binding site itself becomes unstable, rendering the zinc either very loosely bound or else totally free in solution.

The status of the bridging imidazolate hypothesis is not entirely sweetness and light, however, because a serious contention has surfaced recently.[209] In experiments analogous to those cited above for the cobalt-copper and cadmium-copper derivatives of bovine dismutase, the spectrum of a cadmium-only protein was found to be identical to that of the cadmium-cupric enzyme, i.e., the spectroscopic properties of cadmium-only protein were identical to oxidized enzyme. These data, obtained from perturbed angular correlation of gamma ray spectroscopy, using the ^{111}Cd isotope, suggest that cadmium, and by inference zinc, is not liganded to the bridging imidazolate in the resting (oxidized) enzyme. If the His 61 imidazolate did bridge, then the resultant negative charge on the imidazole ring in the cupric state of the protein (Equation 5) would grossly perturb the spectrum of ^{111}Cd. Competition experiments demonstrated that this cadmium isotope was indeed bound to the native zinc site.[209] It is not immediately obvious how to reconcile results from this method with evidence for the presence and catalytic involvement of the bridging imidazolate, cited earlier. The Cu-Zn enzyme used in this conflicting ^{111}Cd study was from bakers' yeast. Species differences have been noted in the superhyperfine EPR spectra of copper and between fungal dismutases and those of higher organisms,[183] although the inclination is to seek explanation elsewhere, in view of the considerable evidence, both chemical and spectroscopic, for evolutionary conservation of active site structure and mechanism in Cu-Zn dismutases.

The identity of the proton donor in the reduction step of dismutation is still a mystery. The proposal that the liberated imidazole N-H donates a proton remains viable.[8,142] Alternatively, the proton donor might be water which is bound in the active site region, or the guanidinium function of bovine Arg 141, which has been catalytically implicated in chemical modification studies of bovine and yeast enzymes (Section III.C).[8,144-147]

6. Assignment of Proton Resonances in NMR Spectra

In the studies discussed previously, NMR spectroscopy was used diagnostically, the spectra being interpreted as composite fingerprints of the different chemical environments of nonexchangeable dismutase protons. Further information can be extracted from these spectra if specific resonances can be assigned to protons of specific residues in the amino acid sequence.

The assignment of resonances as "zinc ligands" or "copper ligands" can be made by comparing spectra of apo, zinc-only and holoenzyme. Different approaches are used for assignments of imidazole N-H protons and imidazole C-2 and C-4 protons, although the strategy for both is based upon the same principle: that the protons of metal ligands will be exchanged at a greatly reduced rate. With imidazole N-H protons, it is exchange with solvent water which is reduced. Thus, a resonance which is enhanced in the spectrum of zinc-only protein, but weak or absent in the spectrum of apoprotein, may be assigned to a histidine which is a ligand of zinc. Similarly, a strong resonance in the spectrum of holoenzyme, which is weak or lacking in that of zinc-only protein, may be assigned to a copper-bound imidazole function. Using this approach for imidazole N-H protons, copper and zinc ligands have been assigned in NMR spectra of Cu-Zn dismutases from bovine erythrocytes and wheat germ (isozymes I and II), and the metal-binding centers of these three enzymes have been shown to be similar.[185,200]

Resonances of imidazole C-2 protons are assigned by slightly different approaches. For nonligand histidine residues, which are exposed to solvent, chemical shifts will ordinarily be dependent upon solution pH, moving upfield with increasing pH (pH 5 to 9). The resonances of C-2 protons of histidine ligands will not show such a titration behavior, and can be identified in ^{1}H NMR spectra of the holoenzyme by their pH-independent chemical shifts.[196] Distinction between zinc ligands and copper ligands is not possible in using this approach. The resonances of five histidine ligands have been identified for bovine erythrocyte[195,196] and human erythrocyte dismutases[196,210] and the similarity in metal-binding site environments of the two enzymes has been deduced from the virtual identity of the corresponding chemical shifts of the C-2 protons.

Bovine and human SOD each contain eight histidine residues per polypeptide chain. Six of these are known to be ligands of copper and zinc, from the crystal structure of the bovine enzyme[85,86] and homologies between the two sequences.[77-81] There are two resonances of C-2 protons which show pH-dependent shifts in the proton NMR spectrum of the bovine holoenzyme.[195] These have been assigned to His 19 and His 41, because these two are known to be nonactive site residues from the crystal structure. Two similar resonances exist in the NMR spectrum of the human holoenzyme.[210] One of these arises from His 43, which is homologous to bovine His 41.[80,81] Bovine His 19 is not conserved in the human dismutase sequence, and the second of the two nonactive site histidine C-2 proton resonances in the human NMR spectrum is consequently assigned to His 110. The imidazole side chain of this residue evidently lies on the surface of the human enzyme,[210] analogous to the bovine His 19 imidazole. However, these two sequence positions (residues 19 and 110) are not particularly close in the three-dimensional structure of the bovine enzyme.

Another approach for assigning C-2 proton resonances, which does allow distinction between copper and zinc ligands, utilizes deuterium exchange and the diminished exchange rate for C-2 protons of histidine ligands. The apoprotein, zinc-only protein, and holoenzyme are incubated in D_2O for 12 days at 50°C., during which time exposed C-2 protons (and many other exposed protons as well) undergo deuterium-for-hydrogen exchange. The apoprotein is then reconstituted with copper and with zinc; the zinc-only protein is reconstituted with copper, and proton NMR spectra are recorded for all three deuterated proteins. The only resonances observed in the ^1H NMR spectrum will be those from protons which exchanged very slowly or not at all.[187] For example, in the spectrum of deuterated holoenzyme, the C-2 resonances of metal ligands will be observed; resonances of surface and/or nonactive site histidine residues will be absent and can be identified by comparison with the spectrum of undeuterated holoenzyme. Similarly, C-2 proton resonances of both nonactive site histidine and copper ligands will be absent from the spectrum of deuterated, then reconstituted, zinc-only protein. Resonances of nonactive site and of all metal ligand histidine residues will be absent from the spectrum of deuterated then reconstituted apoprotein. By suitable comparisons of spectra of deuterated and nondeuterated proteins, assignments of C-2 proton resonances to nonactive site, to copper, and to zinc ligands have been made in the NMR spectra of bovine and yeast dismutases.[187,196] No surface or nonactive site histidine residues are detected in spectra of the yeast enzyme, consistent with sequence comparisons which show only six histidine residues in the polypeptide sequence of yeast dismutase, all of which are homologous to known metal ligands in the bovine structure (Section III.D.2).[77-79,82,83]

One of the nonligand C-2 protons in bovine dismutase undergoes exchange extremely rapidly; its resonance decreases when the spectrum of undeuterated enzyme is recorded in D_2O. Following incubation in tritiated water, rather than D_2O, and analysis of peptic peptides from radiolabeled holoenzyme, histidine 41 has been identified as the exclusive site of tritiation. In this fashion, a specific resonance in the NMR spectrum has been assigned to this specific histidine residue in the sequence.[211] By difference, the resonance of His 19 (or His 110) C-2 proton is then known, as the second of the two nonactive site resonances in the proton NMR spectra.[187,196,210]

Using a combination of approaches, pH dependence, deuterium-for-hydrogen exchange, and effect of perchlorate vs. chloride anions on chemical shifts, resonances of histidine C-2 protons have been assigned and compared in NMR spectra of bovine, human swordfish, and yeast copper-zinc dismutases.[196] As noted earlier, a resonance assigned to freely titrating His 19 and His 110 is identified in spectra of bovine and human enzymes, respectively, and is lacking in the spectrum of yeast enzyme. From the absence of this resonance in the spectrum of swordfish enzyme, the absence of His 19 or His 110 in its as-yet-to-be-determined sequence is predicted. The resonance of bovine His 41 shows a characteristic unusual, pH dependence; a similar resonance in the spectrum of swordfish enzyme predicts a conservation of this histidine residue and of its three-dimensional environment, responsible for the atypical pH behavior. Resonances of two copper ligands and three zinc ligands are, expectedly, visible in NMR spectra of all four dismutases. The chemical shift of the C-2 protons of copper ligands are virtually identical for all four, and indicate a highly conserved binding site structure. However, two of the resonances of zinc ligands in the yeast protein are slightly different in their behavior from those of the three vertebrate species. Thus, while the structure of the zinc-binding site is very strongly conserved among vertebrates, this site has evidently undergone some change since the evolution of yeast. Such detailed comparative knowledge about metal-binding sites, from proton NMR spectra, is indeed impressive, and enormously expands the horizons beyond the conservation of functional groups, deduced from sequence homologies alone (Section III.D.2).

Mention must be made of one final approach, that of assigning resonances of exchangeable imidazole protons through study of solvent exchange and the nuclear Overhauser effect. This approach has been applied to bovine liver Cu-Zn dismutase.[212] In a nuclear Overhauser effect experiment, a specific resonance is saturated, and the spectrum monitored a few milliseconds later. Resonances arising from protons which are less than about 3.5 Å from the saturated resonance are influenced and may be detected by a difference spectrum. For example, if an N-H resonance is saturated, then usually resonances of the C-2 and C-4 protons of the same imidazole group are detected.[212] From prior assignment of the resonance of the His 41 C-2 proton,[211] resonances of the N1 and N3 protons of His 41 have been assigned. The N1 resonance has an unusually high pK_a (probably related to the rapid C-2 proton exchange of this residue)[211] which may arise from interactions of His 41 with Glu 119 and Asp 40. Resonances have also been assigned to the N3 protons of His 44 and His 69, both of which are metal ligands. These two resonances have extremely low solvent exchange rates, which have been attributed to the hydrogen bond which each forms with Asp 122. It has consequently been proposed that this hydrogen bond link between these two histidine ligands may be important in the stabilization of the metal-binding site.[212]

In conclusion, the assignment of resonances in proton NMR spectra is seen to be capable of extremely high structural resolution. The great significance of these assignments is that the spectroscopist now has a collection of probes, of specific regions in the three-dimensional structure of bovine dismutases (e.g., the region around the C-2 of His 41 and the environments of N3 in His 44 and His 69), which may be used to monitor structural changes in that protein or to evaluate the degree of structural conservation in dismutases from other species.

H. Structural Aspects of Isozymes and Genetic Variants

The literature is rife with examples of electrophoretically distinguishable variants of Cu-Zn dismutases, in species spanning the evolutionary gamut from *Neurospora* to man. Using activity staining,[182] multiple bands may be observed after conventional electrophoresis[11,19,22,25,26,33,156,157] or, often with greater resolution, after isoelectric focusing.[213] Little is known about the significance of the isozymes. Likewise, the chemical basis for their electrophoretic differences has not been identified, although it appears that the isozymes are not mere artifacts of the purification procedures. The multiple banding is not an artifact of the McCord-Fridovich organic solvent treatment, because the banding patterns of dismutases purified from chicken liver and from bovine erythrocytes are identical to those in crude liver homogenate and in the hemolysate, respectively.[26,156] The banding pattern of chicken liver enzyme is not altered by treatment with mercaptoethanol.[26] Furthermore, the isozymes of the bovine enzyme are not variants with a less-than-stoichiometric complement of copper and zinc, because the banding pattern is unchanged if apoprotein is reconstituted with excess metals.[156] Finally, the human erythrocyte enzyme is multiple banded on electrophoresis, even when purified by the "chromatographic procedure", which avoids organic solvents and prevents formation of the persulfide function.[30] It is likely that the amino acid sequence analyses, upon dismutases from bovine erythrocytes, human erythrocytes, and horse liver, were performed on a mixture of isozymes, yet none of the published structures indicate any microheterogeneity.[77-81,84] It is conceivable that some variants arise from different degrees of amidation, which might be difficult to detect in sequence studies.

Characterizations of isolated variants seem to have been reported in only two instances. Isozymes I and II from wheat germ have been purified.[19] The two proteins differ in electrophoretic mobility, amino acid composition, stability to sodium dodecyl sulfate, and superhyperfine EPR spectra of copper. Even in a nonhybrid strain of wheat, both are present, although leaves and stalks contain only isozyme II, while both are present in the seed heads. No biological reason has been offered for the presence of isozymes in this species.[19]

A basic electrophoretic variant of human Cu-Zn dismutase was detected in an extensive erythrocyte screening study, initially in the heterozygous[214] and later in the homozygous form.[215] The highest frequencies of the homozygous phenotype, termed SOD-2, were found in northern Sweden and northern Finland, especially in Tornedalen, the river valley between Sweden and Finland. In a population study, screening erythrocyte and placental samples, the variant was detected (as the heterozygous form) only in Caucasian populations, with the highest frequencies in northern Sweden and Finland, and also in the Westray Islands of the Orkneys, which were settled by individuals of Scandinavian descent.[216] This genetic variant has been purified from human erythrocytes by the McCord-Fridovich method.[33] Consistent with its electrophoretic mobility, the variant has an isoelectric point (4.85) which is higher than that of normal dismutase (4.7), although no definitive differences are found between the amino acid compositions of the two.[33] A physiological explanation for the presence of this variant in the human population has not been proposed, and it is uncertain whether its high frequency in Sweden and Finland is a matter of chance or selection pressure.[216] The variant is less stable than normal human erythrocyte dismutase at acid and alkaline pH, and less thermostable at neutral pH, although presence of the variant has not been correlated with any particular disease in its carriers.[33] With the amino acid sequence of normal human dismutase now known,[80,81] identification of the amino acid replacement(s) in the variant would be of some interest.

Using cell hybridization techniques, chromosome assignments have been made for the human genes for cytosolic copper-zinc and for manganese-containing superoxide dismutase (found in liver mitochondria[73,74]): the gene for Cu-Zn dismutase to chromosome 21[217] and the gene for Mn dismutase to chromosome 6.[218] In cases of trisomy 21 (Down's syndrome) the activity of erythrocyte (Cu-Zn) dismutase is increased about 50% above normal, a probable gene-dosage effect.[219]

IV. MANGANESE/IRON SOD

A. Introduction

Manganese- and iron-containing SODs have been intensely studied in investigating the physiological function of dismutases. Prokaryotes, which have proven very amenable in elucidating the nature of oxygen toxicity and the responses to it, contain al-

```
           1                                       10                                          20
B. stear.  Pro-Phe-Glu-Leu-Pro-Ala-Leu-Pro-Tyr-Pro-Tyr-Asp-Ala-Leu-Glu-Pro-His-Ile-Asp-Lys
E. coli    Ser-Tyr-Thr-Leu-Pro-Ser-Leu-Pro-Tyr-Ala-Tyr-Asp-Ala-Leu-Glu-Pro-His-Phe-Asp-Lys

           21                                      30                                          40
B. stear.  Glu-Thr-Met-Asn-Ile-His-His-Thr-Lys-His-His-Asn-Thr-Tyr-Val-Thr-Asn-Leu-Asn-Ala
E. coli    Gln-Thr-Met-Glu-Ile-His-His-Thr-Lys-His-His-Gln-Thr-Tyr-Val-Asn-Asn-Ala-Asn-Ala

           41                                      50                                          60
B. stear.  Ala-Leu-Glu-Gly-His-Pro-Asp-Leu-Gln-Asn-Lys-Ser-Leu-Glu-Glu-Leu-Leu-Ser-Asn-Leu
E. coli    Ala-Leu-Glu-Ser-Leu-Pro-Glu-Phe-Ala-Asn-Leu-Pro-Val-Glu-Glu-Leu-Ile-Thr-Lys-Leu

           61                                      70                                          80
B. stear.  Glu-Ala-Leu-Pro-Glu-Ser-Ile-Arg-Thr-Ala-Val-Arg-Asn-Asn-Gly-Gly-Gly-His-Ala-Asn-
E. coli    Asp-Gln-Leu-Pro-Ala-Asp-Lys-Lys-Thr-Val-Leu-Arg-Asn-Asn-Ala-Gly-Gly-His-Ala-His-

           81                                      90                                          100
B. stear.  His-Ser-Leu-Phe-Trp-Thr-Ile-Leu-Ser-Pro-Asn-Gly-Gly-Gly-Glu-Pro-Thr-Gly-Glu-Leu
E. coli    Asn-Ser-Leu-Phe-Trp-Lys-Gly-Leu-Lys-Lys-Gly-Thr-Thr-Leu-Gln-  -     Gly-Asp-Leu

           101                                     110                                         120
B. stear.  Ala-Asp-Ala-Ile-Asn-Lys-Lys-Phe-Gly-Ser-Phe-Thr-Ala-Phe-Lys-Asp-Glu-Phe-Ser-Lys
E. coli    Lys-Ala-Ala-Ile-Glu-Arg-Asp-Phe-Gly-Ser-Val-Asp-Asn-Phe-Lys-Ala-Glu-Phe-Glu-Lys

           121                                     130                                         140
B. stear.  Ala-Ala-Ala-Gly-Arg-Phe-Gly-Ser-Gly-Trp-Ala-Trp-Leu-Val-Val-Asn-Asn-Gly-Glu-Leu
E. coli    Ala-Ala-Ala-Ser-Arg-Phe-Gly-Ser-Gly-Trp-Ala-Trp-Leu-Val-Leu-Lys-Gly-Asp-Lys-Leu

           141                                     150                                         160
B. stear.  Glu-Ile-Thr-Ser-Thr-Pro-Asn-Gln-Asp-Ser-Pro-Ile-Met-Glu-Gly-Lys-  -   -   -   -
E. coli    Ala-Val-Val-Ser-Thr-Ala-Asn-Gln-Asp-Ser-Pro-Leu-Met-Gly-Glu-Ala-Ile-Ser-Gly-Ala-

           161                                     170                                         180
B. stear.   -   -Thr-Pro-Ile-Leu-Gly-Leu-Asp-Val-Trp-Glu-His-Ala-Tyr-Tyr-Leu-Lys-Tyr-Gln
E. coli    Ser-Gly-Phe-Pro-Ile-Leu-Gly-Leu-Asp-Val-Trp-Glu-His-Ala-Tyr-Tyr-Leu-Lys-Phe-Gln

           181                                     190                                         200
B. stear.  Asn-Arg-Arg-Pro-Glu-Tyr-Ile-Ala-Ala-Phe-Trp-Asn-Val-Val-Asn-Trp-Asp-Glu-Val-Ala
E. coli    Asn-Arg-Arg-Pro-Asp-Tyr-Ile-Lys-Glu-Phe-Trp-Asn-Val-Val-Asn-Trp-Asp-Glu-Ala-Ala

           201
B. stear.  Lys-Arg-Tyr-Ser-Glu-Ala-Lys-Ala-Lys
E. coli    Ala-Arg-Phe-Ala-Ala-Lys-Lys
```

FIGURE 5. Complete amino acid sequences of the manganese-containing SODs from *Bacillus stearothermophilus* (above), and *Escherichia coli* B (below). The residue positions are numbered consecutively, including the positions of gaps. Identical residues in the two sequences are enclosed in boxes.[44,89] (Figure 5B reprinted with permission from Brock, C. J. and Walker, J. E., *Biochemistry,* 19, 2873, 1980. Copyright 1980, American Chemical Society.)

most exclusively enzymes of this family (for exceptions, see Section II.D.3). In certain eukaryote cells, which contain both Cu-Zn and Mn forms, it is the content of the manganese rather than the copper-zinc dismutase, which is increased in response to oxygen toxicity threats (see Chapters 4 and 5 of this volume). Detailed knowledge about the structure and mechanism of Mn/Fe enzymes has not grown apace with their physiological studies, and is neither as rich nor as refined as that for the Cu-Zn enzymes (Section III). The lesser stability of the Mn/Fe enzymes, their higher molecular weight, and the greater difficulty in their purification are all probable reasons for their being given short shrift in structure-function studies. Furthermore, the propitious bimetallic active site of the Cu-Zn dismutases, which bounteously permits noncatalytic zinc to be replaced by a spectroscopic probe with retention of catalytic activity (Section III.G), is lacking in the Mn/Fe dismutases. Each of the Mn/Fe enzymes contains but a single type of metal co-factor and no known metal replacement supports catalysis (Section IV.D.2). Finally, chemical modifications and complete amino acid sequence determinations have been few, and no three-dimensional structure has yet been reported for any member of the Mn/Fe dismutase family. Several crystallographic projects have been initiated (manganese enzymes from *S. cerevisiae, E. coli* B,[220] *B. stearothermophilus*,[221,222] and iron enzymes from *P. ovalis*[223,224] and *E. coli*[225]), but none has reached fruition yet. However, a structure of the *P. ovalis* (Fe) dismutase seems imminent;[224] its report would contribute much to resolving questions about metal stoichiometry and active site structure, and might provide a stimulation for much needed chemical and spectroscopic characterizations of this dismutase family.

B. Amino Acid Sequence Analysis

Comparative amino terminal sequence data for Mn/Fe dismutases have been presented and discussed earlier (Section II.E, Figure 2). While such partial sequence data are known for 15 Mn/Fe SODs, including both manganese and iron, prokaryote and mitochondrial forms, and enzymes from both aerobic and obligately anaerobic organisms, complete primary structures have been reported for only two enzymes; both are dimeric manganese-containing dismutases from bacteria (Figure 5).[44,89,102] The complete sequence from the thermophile, *B. stearothermophilus*, has a polypeptide chain of 203 residues, corresponding to a molecular weight of 22,740.[44,102] The sequence of the *E. coli* B dismutase contains 205 residues (molecular weight 22,000).[89] It is not surprising that these two sequences are quite homologous; the identity is 60% for the alignment shown in Figure 5.[44] What is surprising is that this percent identity is only marginally larger than the 55% identity between sequences of Cu-Zn dismutases from bakers' yeast and bovine erythrocytes,[82,83] which span a considerably greater phylogenetic distance than the two bacterial species. Is the amino acid sequence conservation in the Mn/Fe family much less than that in the Cu-Zn family, or is the variation among Mn/Fe dismutases merely exaggerated among the bacterial enzymes? The answer to this question requires more complete sequences, particularly of eukaryote Mn/Fe enzymes. Increased sequence variation among bacterial forms has been noted previously: sequences of the cytochromes c_2, from *Rhodospirillacea* alone, vary more than mitochondrial cytochromes *c*, from the whole span of eukaryote species.[121,122,226] When the complete sequences of these two manganese dismutases were compared with the complete sequences of Cu-Zn dismutases known at the time, no significant homology was detected.[44,89] Thus prior inferences based solely upon comparisons of amino terminal segments were confirmed:[121,122,227-229] The Cu-Zn and Mn/Fe families of SODs have evolved independently; there is no sequence evidence for a common ancestral dismutase protein, from which both present day Mn/Fe and present day Cu-Zn SODs have evolved (Section II.E).

Secondary structures have been predicted from the sequence of the *B. stearother-*

mophilus and the *E. coli* B dismutases by a statistical method.[230] The structures of the two dismutases are very similar to one another, and are uncontestably different from that of the bovine erythrocyte Cu-Zn dismutase.[44,122] Predicted contents of α-helix for these two manganese enzymes are 45 and 61%, for the *B. stearothermophilus* and *E. coli* proteins; respective contents of β-sheet are 34 and 27%. The secondary structure content of the bovine erythrocyte Cu-Zn enzyme was reasonably predicted by this statistical method: observed (predicted), α-helix 3 to 8% (11%) and β-sheet 45% (61%). The independent evolution of Mn/Fe and Cu-Zn dismutase families may have proceeded in a parallel or a convergent fashion (Section II.E). These secondary structure predictions do not absolutely eliminate the possibility of convergent evolution; a crystal structure of a manganese or iron enzyme might reveal a highly homologous catalytic site structure and mechanism, ensconced within a protein conformation grossly different from that of Cu-Zn dismutases.[125]

In the alignment of *B. stearothermophilus* and *E. coli* sequences (Figure 5), seven histidine residues occur at homologous positions (residues 17, 26, 27, 30, 31, 78, and 167, in the *B. stearothermophilus* numbering). A particularly striking cluster occurs in the identical sequence, from residues 26 to 31: His-His-Thr-Lys-His-His. The histidine at residues 30 and 31 is invariant in all known amino terminal sequences of Mn/Fe dismutases (Figure 2), and unpublished NMR studies (cited in Reference 44) have implicated histidine residues as metal ligands in the *B. stearothermophilus* dismutase. It may be that these two conserved histidine residues are ligands of manganese or iron in this dismutase family. It is not known if the two homologous histidines, which occur later in the sequences of the *B. stearothermophilus* and *E. coli* proteins, are conserved in other manganese and iron enzymes.

C. Immunology, Hybrids, and Isozymes

Not unexpectedly, enzymes of the Mn/Fe family are immunologically distinct from Cu-Zn dismutases, in comparisons of both Mn and Fe enzymes, and in comparisons both within and between species.[11,53,54,56,64,231] Even among themselves, however, Mn/Fe dismutases exhibit a high degree of immunological eclecticism. Antibody to the iron-containing dismutase from *Plectonema boryanum* does not react with the iron dismutases from *Chlorobium thiosulphatophilum*,[37] *Chromatium vinosum*,[38] or *Euglena*.[65] The two manganese-containing dismutases from *Pleurotus olearius* are immunologically distinct.[67] The manganese-containing dismutase from *Porphyridium cruentum* does not cross-react with antibodies to either manganese- or iron-containing enzymes from *E. coli*.[64]

The dismutases from certain bacterial species show unusual immunological specificity.For example, antibody against the manganese-containing dismutase from *Mycobacterium*, strain Takeo, forms precipitin lines with extracts from other *Mycobacteria* species and *Gordena* and *Nocardia* species, but not with extracts from *Pseudomonas aeruginosa, Achromobacter xylosoxidans*, or *Acinetobacter anitratus*.[53] The manganese dismutase from *Streptococcus faecalis* is even more antigenically discriminating, because its antiserum shows cross-reaction only with another *S. faecalis* strain, but not with strains of *S. pyogenes, S. pneumoniae, S. lactis, S. aureus*, or *S. epidermidis*, nor with selected strains of the genera *Micrococcus, Bacillus, Escherichia*, or *Pseudomonas*, (based upon Ouchterlony double diffusion and inhibition of dismutase activity in extracts).[60] While this immunological specificity may decrease the utility of SOD as an antigen for plotting bacterial phylogeny, it has been useful for demonstrating a close relationship between certain species (e.g., among the luminescent bacteria of the genus *Beneckea* and *Photobacterium* [232]). This immunological diversity among bacterial manganese dismutases is consistent with the relatively small degree of sequence homology between manganese enzymes from *Bacillus* and *Escherichia* species[44,102] (Section IV.B).

Interspecies relatedness among Cu-Zn dismutases is greater when measured by hybridization studies, reflecting subunit interaction forces, than when measured by immunological methods, reflecting surface properties (Section III.E). There appear to be no reports of in vitro hybridization studies with Mn/Fe dismutases, although such investigations would be of interest in view of (1) results with Cu-Zn enzymes, (2) the low degree of Mn/Fe immunological similarity, and (3) existence of both dimeric and tetrameric manganese-containing enzymes. A naturally occurring hybrid does exist; aerobically grown *E. coli* B contain (in addition to an iron- and a manganese-containing dismutase) a hybrid which has been purified to homogeneity.[50] The hybrid is a dimer, containing one Mn dismutase type of polypeptide chain and one Fe dismutase type of polypeptide chain. Its specific activity is comparable to that of the other *E. coli* dismutases, and it contains only iron, 0.8 g-atoms/mol of dimer. The existence of this hybrid is proof positive of homology in the structure and forces at the subunit interfaces of the Mn and Fe dismutases from *E. coli.* The presence of iron rather than manganese in the hybrid may be more a consequence in vivo of metal availability than metal-binding specificity of the active site.[50] It would seem to be particularly interesting to utilize this hybrid in in vitro studies of metal replacement and active site mechanism, such as have been performed on homopolypeptide dismutases (Section IV.D.2). Hybridization was not detected between the manganese-containing dismutase from *S. faecalis* and the bovine erythrocyte Cu-Zn enzyme, using the urea dissociation technique.[156]

Isozymes of manganese- and iron-containing dismutases have been detected in several species using electrophoresis or isoelectric focusing,[11,40,114,213] and have actually been purified from *S. mutans,*[61] *Euglena gracilis,*[65] and bovine heart mitochondria.[72] Predictably, the isozymic species seem to have identical molecular weight, subunit structure, and metal content, and to differ only in their charge properties. The exact chemical basis for the charge difference has not been elucidated in any instance, and no biological explanation has been forwarded for the existence of isozymic forms in any species. In bacterial species which contain both Mn and Fe dismutases, a hybrid band is often seen following electrophoresis;[114] this might be mistaken for an isozymic form. Potential artifactual sources of electrophoretic or isoelectric heterogeneity have also been noted. (1) Streptomycin sulfate, routinely used for precipitation of nucleic acids in the purification of many bacterial enzymes, shows a selective reaction with dismutases of certain bacterial species; other bacterial dismutases are unreactive, as are Cu-Zn dismutases.[233] The reaction is covalent, appears to involve the formyl group of streptomycin, and alters the electrophoretic mobility without change in activity or molecular weight of the dismutase. (2) The isozymic forms of bovine heart mitochondrial Mn dismutase may bind ampholytes, producing a profusion of bands on isoelectric focusing.[72]

D. Catalytic Mechanism and Structure of the Active Site

1. Introduction

Spectroscopic studies of metal replacement derivatives have been particularly important in mechanistic analysis of Cu-Zn dismutases. Substitution of Co^{2+} or $^{113}Cd^{2+}$ for Zn^{2+}, in copper-containing enzyme, allows for monitoring active site structure of a catalytically active species and changes in metal environment during the oxidation and reduction steps of catalysis. Such an opportunity has been precluded in Mn/Fe dismutases by the presence of but one type of metal in the active site. Despite this logistical difference, metal replacement studies have been a fruitful approach, which, in combination with spectroscopic methods and pulse radiolysis kinetic techniques, has been an important vehicle of insight into active site structure and mechanism of the Mn/Fe enzymes.

2. Metal Reconstitution Studies

The viability of metal replacement has been demonstrated with dimeric manganese-containing dismutases from the bacteria, *B. stearothermophilus*[43,101,102] and *E. coli* B,[103,104] dimeric iron-containing dismutases from the bacteria, *Pseudomonas ovalis*[224,234-236] *and Photobacterium leiognathi*,[237] the tetrameric manganese-containing enzyme from the bacterium, *Thermus aquaticus*,[62] and the dimeric manganese-containing enzyme from the eukaryote alga, *Porphyridium cruentum*.[64] In all instances, the inactive apoprotein has been reconstituted with the native metal ion, yielding an active enzyme of the same metal stoichiometry as native holoenzyme. The specific activity of reconstituted enzyme was about 50% for the *P. leiognathi* Fe dismutase,[237] but close to 100% of the value for native enzyme in all other instances.

Reconstitution of apoprotein with a non-native divalent metal ion has been performed with four of these dismutases, those from *B. stearothermophilus, E. coli* B, *P. leiognathi*, and *P. ovalis*. It has been possible in all four cases to replace the native manganese by iron, or replace the native iron by manganese ion.[101,102,104,224,236,237] Substitutions of other non-native metal ions have also been made, with the different enzymes showing slightly different binding specificities. For example, the *E. coli* (Mn) apoprotein has been reconstituted with cobalt, nickel, and zinc, but not with copper,[104] while the *B. stearothermophilus* (Mn) apoprotein does reconstitute with copper, as well as cobalt and nickel.[101,102] The *P. leiognathi* (Fe) apoprotein has been reconstituted with nickel, copper, chromium, cobalt, and zinc,[237] while the *P. ovalis* (Fe) apoprotein binds neither nickel nor copper, but does reconstitute with chromium and cadmium.[224,236]

Despite these differences in metal-binding specificity, the metal replacement derivatives of Mn/Fe dismutases share a number of common properties (1) In most instances, replacement metals (both native and non-native) bind with a stoichiometry close to 1 cation per dimer, or about 0.5 metal per subunit.[43,62,64,102-104,224,234-237] (2) This stoichiometry and the competition with the native manganese (shown only in the *E. coli* enzyme studies)[103,104] suggest that the replacement metal occupies the native metal-binding site. (3) Only the divalent forms of the cations bind to the apoproteins. For example, neither Fe^{3+} nor Cr^{3+} reconstitute with the *P. leiognathi* apoprotein.[237] (4) Only the native manganese or iron supports catalytic activity. All metal replacements (even manganese-for-iron and iron-for-manganese) are inactive.[101-104,224,236,237]

In the resting state of the native holoenzyme, manganese and iron are present in the tervalent state, Mn^{3+}[47] and Fe^{3+}.[39,40,48,49,54-56,59] Although divalent metal ions are added in metal replacement of manganese-for-iron and iron-for-manganese, oxidation to the tervalent state appears to occur during or subsequent to binding of the metal by the dismutase apoprotein. Thus the iron derivative of apoprotein from *B. stearothermophilus* (Mn) dismutase,[101,102] and the manganese derivative of the apoprotein from *P. ovalis* (Fe) dismutase[224,236] have absorption spectra in the visible region which are, respectively. similar to those of native iron- and native manganese-containing dismutases. (The similarities are qualitative; the extinction coefficients may be smaller in the metal replacement proteins than in the native metal proteins.)[236] Furthermore, the presence of ferric iron has been confirmed by *o*-phenanthroline treatment of *B. stearothermophilus* (Mn) apoprotein, reconstituted with ferrous ion.[102] These properties of the Mn-for-Fe and Fe-for-Mn replacement derivatives lend themselves to some rather fascinating evolutionary proposals. That Mn/Fe dismutases have evolved from a common ancestral dismutase is attested to by the strong homologies in their amino terminal sequences (Section II.E.). However divergence has not been excessive, and from our current evolutionary vantage point, we still find functional vestiges of their common heritage. Thus the structures of the catalytic sites of Mn dismutases and Fe dismutases still bear sufficient homology to bind the divalent metal co-factor of the other enzyme form and to oxidize it from divalent to tervalent state, ie., the valence change of one half of the catalytic mechanism (Section IV.D.5). The metal replacement derivatives are not catalytically competent; evidently the divergence has been extensive enough to preclude this once-shared property.[101,102]

The conditions for removing metals from Mn/Fe dismutases appear to dissociate the subunits of the multimer. However, all proteins, reconstituted with either native non-native metals, and the apoprotein forms as well, seem to regain the dimeric or tetrameric structure characteristic of the particular parental holoenzyme.[43,62,102] These observations indicate that while the metal ion is indispensable for catalysis, it is hardly a *sine qua non* for subunit association. CD spectra and the reactivity of sulfhydryl groups towards 5,5′-dithiobis (2-nitrobenzoic acid) have been presented as further evidence for the conformational similarity of native *P. ovalis* (Fe) dismutase and its metal replacement derivatives.[224]

Detailed knowledge about the architecture of the metal-binding sites is not available for Mn/Fe dismutases, although the above data do circumstantially suggest that metal binding may occur between two subunits, the site being formed in a subunit contact region.[43,102] This suggestion is based upon (1) the stoichiometry of about 0.5 metal ion per polypeptide chain, or 1.0 ion per two polypeptide chains, for reconstituted dimeric and tetrameric dismutases and (2) the need of subunit-dissociating conditions for either forming the apoprotein from native enzyme or for reincorporating metal into the apoprotein.[64] Should this proposed location prove to be correct, then the basic structure of the metal-binding site would be quite different from that in Cu-Zn dismutases (Section III.B.1), where the catalytic site is near the exterior of the molecule, distant from the subunit interface.[85,86] Such a difference in the solvent access to the metal site might account for the fact that polyacrylamide gel electrophoresis is able to separate holoenzyme apoprotein, and partially reconstituted derivatives of bovine Cu-Zn dismutase,[173] while the holoenzymes and apoproteins of Mn/Fe dismutases have identical electrophoretic mobilities.[102,103]

3. Ambiguities about Metal Stoichiometry

The stoichiometry of copper and zinc in Cu-Zn dismutases has been found to be nearly exactly 1.0 of each metal ion per polypeptide chain in a great many preparations of these enzymes from a great many different species. (Table 1) Such consistent contents have not been found in the Mn/Fe family, where differences in the ratio of metal-to-polypeptide chain have been different for enzymes of different species (Table 2). Different metal contents, sometimes varying by a factor of 2, have also been reported for the same dismutase, isolated from the same species, by two different laboratories (*E. coli* [Mn],[45,47] *E. coli* [Fe],[48,49] and *P. boryanum* [Fe] enzymes),[56,57] and in successive preparations from the same laboratory (*P. leiognathi* [Fe][99,237] and *P. ovalis* [Fe][236] enzymes). The values have ranged from about 1 g-atom of Mn or Fe per mole of polypeptide subunit to about 0.5 g-atom/mol of subunit. These lower, substoichiometric values have long been a source of consternation to dismutase enzymologists and have often been attributed to loss of metals during enzyme purification.[5] However, accumulated data from two sources suggest that metal contents of 0.5 g-atom/mol of subunit are of functional significance and ought not be so speedily dispensed as purification artifacts.

One source of data is the reconstitution studies cited above (Section IV.D.2), in which values of approximately 0.5 g-atom of metal per mole of subunit have been reported for apoproteins reconstituted with either native or non-native divalent metal ions. Considering that the apoproteins of Mn/Fe dismutases are customarily exposed to an enormous molar excess of metal ions during reconstitution and that the treatments for removing nonspecifically bound metal have been both relatively rapid and relatively mild, compared to an enzyme purification scheme, values of 0.5 cation per polypeptide do not seem so easily ascribed to artifactual losses. A second and perhaps more convincing body of data is that which demonstrates no increase in enzyme activity when metal contents are increased from 0.5 to 1.0 g-atom/mol of polypeptide chain. The same specific activity is exhibited by different preparations of *P. ovalis* dismutase, reconstituted with native iron to different extents (1.1 to 1.3 g-atoms/mol of dimer);[235] a comparable range is found for different preparations of the native enzyme (1.1 to 1.5 g-atoms/mol of dimer).[236] Different preparations of the *P. leiognathi* (Fe) dismutase have iron contents of 1.1 to 1.5 g-atom/mol of dimer, yet have the same activity both by conventional assay methods and pulse radiolysis technology.[99,237] Finally, if the apoprotein of *E. coli* (Mn) dismutase is reconstituted with increasing amounts of Mn^{2+}, the activity increases; the plateau at the activity of the native enzyme occurs at 0.5 g-atom of manganese added per mole of subunit.[104]

The above evidence seems rather convincing, that 0.5 metal ion per subunit polypeptide is sufficient for maximal activity. However, for the *P. ovalis* dismutase, it has been posited that there actually may be two metal-binding sites and that the observed content of one metal ion per dimer is the consequence of negative cooperativity between the two subunits.[224] These crucial issues, the number of catalytic sites and the number and location of the metal-binding sites, will ultimately be resolved by three-dimensional structural data on Mn/Fe dismutases.

An accurate value for metal content depends upon both the method of metal analysis and the method of protein determination. A systematic overestimation of the manganese content of dismutases arises from using the method of Murphy and Kies[238] to determine protein concentration. The spectrum of manganese dismutases shows an absorbance shoulder near 225 nm, which is not apparent in the spectrum of other dismutases.[26,73] Thus the published value for the manganese content of the *E. coli* B (Mn) dismutase, 1.8 g-atoms/mol of dimer, has been revised downward to 1.2 g-atoms/mol of dimer.[46]

4. Chemical Modification Studies

Chemical modification of Mn/Fe dismutases, towards the end of identifying functional or metal-binding amino acid residues, is a relatively unexplored domain. The iron enzyme from *P. leiognathi* appears to have been studied most extensively. In one study, tryptophan was implicated as a catalytically involved residue.[55] When the holoenzyme was inactivated by irradiation, there was a parallel decrease in tryptophan fluorescence, and a blue shift in the tryptophan fluorescence maximum. However, this implication was regarded as tentative, because the possibility of chemical modification of other photosensitive residues was not evaluated, and because loss of activity occurred 2.6 times faster than the decrease in fluorescence.[55] In another study,[239,240] using a larger battery of reagents, both histidine and tryptophan were implicated as catalytic amino acids, because the holoenzyme was inactivated during modification by diethylpyrocarbonate and 2-hydroxy-5-nitrobenzyl bromide, respectively. As the authors note, the surface charge and hydrophobicity of the holoenzyme might well be altered by reaction with these two reagents, and the strength of the catalytic implication depends upon eliminating the possibility of conformational changes in the holoenzyme as a cause of inactivation. (Chemical modification of dismutases may result in loss of the metal co-factor;[144] this loss may be the origin of the inactivation. Determining the metal content of chemically modified derivatives is consequently of some importance.) When the reactivities of the apoprotein and of the *P. leiognathi* holoenzyme, to 2-hydroxy-5-nitrobenzyl bromide and to tetranitromethane, were compared, one additional residue of tryptophan and tyrosine, respectively, were reactive per subunit of the apoprotein. This difference implicated these types of two residues as metal ligands.[239,240] The strength of this proposal is also contingent upon examining the protein derivatives for conformational changes. If the metal-binding site is indeed located at a subunit interface (as discussed in Section IV.D.2), then chemical modification of metal-binding residues might well destabilize subunit interactions, and consequently alter the conformation of the molecule. Another concern is that only 35% of the native holoenzyme activity is regained when the *P. leiognathi* apoprotein is reconstituted with iron.[239,240] This low value suggests that the apoprotein form may not be particularly stable, or that its conformation may differ significantly from that of holoenzyme, in either of which cases, chemically reactive reagents might modify amino acid side chains other than those in the metal binding site region.

5. Pulse Radiolysis

Applications of pulse radiolysis have utilized the same strategy that was employed in examining the kinetic mechanism of Cu-Zn dismutases (Section III.F.1). The rate constant for enzyme-catalyzed decomposition of superoxide is determined from monitoring the pulsed reaction mixture at a wavelength in the UV. Absorbance changes at a visible wavelength, within the absorption peak of the metal co-factor, are used to calculate constants for the rates at which the metal is oxidized and reduced during catalysis.

In pulse radiolysis studies of the iron-containing dismutase from *P. leiognathi*,[99] the second-order rate constant for enzyme-catalyzed decomposition of superoxide radical (5.4×10^8 M^{-1} sec^{-1}, at pH 7.0) was shown to be identical to the second-order rate constants for reduction of the oxidized enzyme and for oxidation of the reduced enzyme. (The metal absorbance was monitored at 350 nm.) These data support a simple two-step mechanism, in which the catalytic iron co-factor is alternately reduced and reoxidized, in exact analogy to the case with Cu-Zn dismutases, in which copper participates in a cyclic reduction-reoxidation mechanism (Section III.F.1). The *P. leiognathi* enzyme does, however, display a bizarre trait: when oxidized enzyme is reduced, the net absorbance change at 350 nm does not equal the absorbance change observed on oxidation of the reduced enzyme.[99] This result smacks of an equally bizarre trait, once ascribed to bovine Cu-Zn superoxide dismutase: the inequality of oxidized and reduced forms of the enzyme under steady state turnover conditions (Section III.F.3).[163,172] While negative cooperativity and half-of-the-sites reactivity were initially rallied in the case of the Cu-Zn enzyme, such was not the case here, because of additional data regarding the *P. leiognathi* (Fe) dismutase.[99] The first datum was that the absorbance of resting enzyme at 350 nm was only partially bleached upon treatment with hydrogen peroxide. Secondly, the second-order rate constant for decay of superoxide anion was identical for preparations of the enzyme with iron contents ranging from 1.15 to 1.53 g-atoms of metal per mole of dimeric protein. On the basis of these two additional facts, an alternative explanation was proposed: that preparations of this enzyme contain a variable amount of noncatalytic, bound iron, that this iron resists bleaching by hydrogen peroxide, but can participate in electron transfer, albeit by a noncatalytic mechanism. However, such arguments based upon kinetics cannot totally exclude the possibility of a conformational basis to the behavior, for example, that the inequality in the absorbance changes arises from a conformational difference between the oxidized and the reduced states of the iron enzyme.[99]

The oxidized ("resting") and reduced states of iron-containing SODs correspond to ferric and ferrous iron, respectively. This assignment is based upon the EPR spectra of iron enzymes from *P. leiognathi*[99] and numerous other prokaryote sources[39,40,48,49,54-56,59] which unambiguously indicate that the resting, or oxidized, enzyme contains ferric iron. In all instances, these spectra display the characteristic cachet of ferric, high-spin iron, in a rhombic environment. The symmetry of the rhombic state seems to be slightly different in the iron enzymes of different species.

The kinetic mechanism of manganese SODs shows an idiosyncratic complication unique unto themselves. In pulse radiolysis studies of *E. coli* B (Mn)[46] and *B. stearothermophilus* (Mn) dismutases,[241] the simple two-step mechanism is operative at low substrate-to-enzyme ratios ($[O_2^-]/[E] < 10$ to 15). The enzyme is alternately converted between oxidized (resting enzyme) and reduced (hydrogen peroxide-treated enzyme) forms. The second-order rate constant for oxidation or reduction is equal to that for the enzyme-catalyzed decomposition of superoxide (1 to 2×10^9 M^{-1} sec^{-1}, at pH 7 to 8). The complication arises at larger excesses of superoxide over enzyme, when additional reactions, involving a third form of the enzyme, become significant. This third form is either a considerably less effective catalytic agent,[46] or is a totally inactive form of the enzyme which can, however, regenerate active, oxidized enzyme.[241] At high substrate to enzyme ratios, the two-step mechanism no longer suffices, and a four-step mechanism is required to describe the interconversions of the three forms. The chemical nature of the third form of the enzyme is not known; it might contain manganese in an oxidation state different from that of oxidized or reduced enzyme.[46]

The actual valence of manganese in manganese-containing dismutases was initially unknown.[45] At sufficiently high-protein concentrations, solutions of resting (oxidized) manganese dismutases are pink and are EPR silent. After heating in 0.1 *N* HCl, the EPR spectrum shows the characteristics of divalent manganese.[45,51,52,61,66,67] The resting enzyme was presumed to contain trivalent manganese, Mn^{3+}, based upon its visible absorption[45,47,100,242] and CD spectra.[242] (Trivalent manganese, which is not a paramagnetic species, is evidently reduced to Mn^{2+} during the denaturation of native enzyme by 0.1 *N* HCl.) This valence assignment was proved definitively by magnetic susceptibility measurements.[47] The simple two-step mechanism of catalysis consequently involves interconversions between Mn^{3+} (resting or oxidized enzyme) and Mn^{2+} (reduced enzyme).

6. Spectroscopic Studies of Catalytic Mechanism

Absorbance, EPR, and NMR spectroscopy have been used to distinguish between an inner and an outer sphere mechanism in catalysis, i.e., whether superoxide anion is bound to the inner or the outer coordination sphere, of the metal, prior to or during electron transfer. NMR measurements of the proton relaxation rates of water have been presented as evidence for an outer sphere mechanism in both the manganese- and the iron-containing dismutases from *E. coli* B.[243] The data show that the exchange rate for metal-bound water is too low, by some two orders of magnitude, to be catalytically important. In addition, the EPR spectrum of the *E. coli* (Fe) dismutase is not altered by cyanide anion, which indicates that cyanide, and presumably superoxide, does not bind to the inner coordination sphere of the metal in the iron enzyme.[244]

The outer sphere mechanism proposed for the iron dismutase has been contested.[49,168] (The iron-containing dismutase used for this later study was prepared from *E. coli* B by a new procedure, which yielded enzyme of higher specific activity and higher metal content, 1.8 vs. the earlier value[48] of 1.0 g-atoms of iron per dimeric molecule.) Although the absence of cyanide binding was confirmed, both optical and EPR titrations proved that azide and fluoride do bind, at one and at two coordination positions per iron atom, respectively. The demonstrated binding of these anions suggests that the inner coordination sphere of iron may be accessible to superoxide anion and that an inner sphere mechanism of electron transfer should remain a viable possibility for iron-containing SODs.[49,168]

The lesser stability of Mn/Fe dismutases may limit the application of certain spectroscopic approaches used so famously with the Cu-Zn enzymes, such as the deuterium exchange method of identifying metal ligands, which requires prolonged incubation at 50°C.[187] Nonetheless, one can quite confidently predict that the combination of spectroscopic technology with a three-dimensional crystal structure will resolve many existing questions and fill the present lack of detailed knowledge about the active site structure and mechanism of manganese-iron SODs.

ACKNOWLEDGMENTS

Financial support during the preparation of this chapter was obtained from the Sinsheimer Foundation and from grant GM24527 of the National Institutes of Health. The author expresses his thanks to Dr. Philip Aisen (Department of Biophysics) and Dr. Jack Peisach (Department of Molecular Pharmacology) of Albert Einstein College of Medicine for discussions about spectroscopic methods, and to Dr. Byron Rubin of the Department of Chemistry at Emory University for a prior gift of a wire polypeptide backbone model of the bovine erythrocyte SOD dimer.

REFERENCES

1. **McCord, J. M. and Fridovich, I.**, Superoxide dismutase. An enzymatic function for erythrocuprein (hemocuprein), *J. Biol. Chem.*, 244, 6049, 1969.
2. **Asada, K., Kanematsu, S., Okaka, S., and Hayakawa, T.**, Phylogenetic distribution of three types of superoxide dismutase in organisms and in cell organelles, in *Chemical and Biochemical Aspects of Superoxide and Superoxide Dismutase*, Bannister, J. V. and Hill, H. A. O., Eds., Elsevier/North-Holland, New York, 1980, 136.
3. **Bannister, J. V. and Hill, H. A. O., Eds.**, *Chemical and Biochemical Aspects of Superoxide and Superoxide Dismutase*, Elsevier/North-Holland, New York, 1980.
4. **Fridovich, I.**, The biology of oxygen radicals, *Science*, 201, 875, 1978.
5. **Fridovich, I.**, Superoxide and superoxide dismutases, in *Advances in Inorganic Biochemistry*, Eichhorn, G. L. and Marzilli, L. G., Eds., Elsevier/North-Holland, New York, 1979, 67.
6. **Hill, H. A. O.**, The superoxide ion and the toxicity of molecular oxygen, in *New Trends in Bio-Inorganic Chemisry*, Williams, R. J. P. and DaSilva, J. R. R. F., Eds., Academic Press, New York, 1978, 173.
7. **Michelson, A. M., McCord, J. M., and Fridovich, I., Eds.**, *Superoxide and Superoxide Dismutases*, Academic Press, London, 1977.
8. **Rotilio, G., Rigo, A., and Calabrese, L.**, Recent developments on the active site structure and mechanism of bovine copper- and zinc-containing superoxide dismutase, in *Frontiers in Physicochemical Biology*, Pullman, B., Ed., Academic Press, New York, 1978, 357.
9. **Vignais, P. M., Henry, M.-F., Terech, A., and Chabert, J.**, Production of superoxide anion and superoxide dismutases in *P. denitrificans*, in *Chemical and Biochemical Aspects of Superoxide and Superoxide Dismutase*, Bannister, J. V. and Hill, H. A. O., Eds., Elsevier/North-Holland, New York, 1980, 154.
10. **Puget, K. and Michelson, A. M.**, Isolation of a new copper-containing superoxide dismutase, bacteriocuprein, *Biochem. Biophys. Res. Commun.*, 58, 830, 1974.
11. **Martin, J. P., Jr. and Fridovich, I.**, Evidence for a natural gene transfer from the ponyfish to its bioluminescent bacterial symbiont, *Photobacter leiognathi*: the close relationship between bacteriocuprein and the copper-zinc superoxide dismutase of teleost fishes, *J. Biol. Chem.*, 256, 6080, 1981.
12. **Rapp, U., Adams, W. C., and Miller, R. W.**, Purification of superoxide dismutase from fungi and characterization of the reaction of the enzyme with catechols by electron spin resonance spectroscopy, *Can. J. Biochem.*, 51, 158, 1973.
13. **Misra, H. P. and Fridovich, I.**, The purification and properties of superoxide dismutase from *Neurospora crassa*, *J. Biol. Chem.*, 247, 3410, 1972.
14. **Goscin, S. A. and Fridovich, I.**, The purification and properties of superoxide dismutase from *Saccharomyces cerevisiae*, *Biochim. Biophys. Acta*, 289, 276, 1972.
15. **Weser, U., Fretzdorf, A., and Prinz, R.**, Superoxide dismutase in baker's yeast, *FEBS Lett.*, 27, 267, 1972.
16. **Weser, U., Prinz, R., Schallies, A., Fretzdorff, A., Krauss, P., Voelter, W., and Voetsch, W.**, Microbial and hepatic cuprein (superoxide dismutase), *Hoppe-Seyler's Z. Physiol. Chemie*, 353, 1821, 1972.

17. **Sawada, Y., Ohyama, T., and Tamazaki, I.**, Preparation and physicochemical properties of green pea superoxide dismutase, *Biochim. Biophys. Acta*, 268, 305, 1972.
18. **Asada, K., Urano, M., and Takehashi, M.**, Subcellular location of superoxide dismutase in spinach leaves and preparation and properties of crystalline spinach superoxide dismutase, *Eur. J. Biochem.*, 36, 257, 1973.
19. **Beauchamp, C. O. and Fridovich, I.**, Isozymes of superoxide dismutase from wheat germ, *Biochim. Biophys. Acta*, 317, 50, 1973.
20. **Bannister, J. V., Anastasi, A., and Bannister, W. H.**, Cytosol superoxide dismutase from swordfish (*Xiphias gladius* L.) liver, *Comp. Biochem. Physiol.*, 568, 235, 1977.
21. **Mennella, M. R. F. and Jones, R.**, Properties of spermatozoal superoxide dismutase and lack of involvement of superoxides in metal-ion-catalyzed lipid-peroxidation reactions in semen, *Biochem. J.*, 191, 289, 1980.
22. **Sharoyan, S. F., Shalijain, A. A., Nalbandyan, R. M., and Buniatian, H. C.**, Two copper-containing proteins from white and gray matter of brain, *Biochim. Biophys. Acta*, 493, 478, 1977.
23. **Bannister, J. V., Bannister, W., and Wood, E.**, Bovine erythrocyte copper-zinc protein. I. Isolation and general characterization, *Eur. J. Biochem.*, 18, 178, 1971.
24. **Keele, B. B., McCord, J. M., and Fridovich, I.**, Further characterization of bovine superoxide dismutase and its isolation from bovine heart, *J. Biol. Chem.*, 246, 2875, 1971.
25. **Magnes, L. J. and Li, T. -K.**, Isolation and properties of superoxide dismutase from bovine spermatozoa, *Biol. Reprod.*, 22, 965, 1980.
26. **Weisiger, R. A. and Fridovich, I.**, Superoxide dismutase. Organelle specificity, *J. Biol. Chem.*, 248, 3582, 1973.
27. **Albergoni, V. and Cassini, A.**, A cupro-zinc protein with superoxide dismutase activity from horse liver. Isolation and properties, *Comp. Biochem. Physiol.*, 478, 767, 1974.
28. **Carrico, R. J. and Deutsch, H. F.**, Isolation of human hepatocuprein and cerebrocuprein. Their identity with erythrocuprein, *J. Biol. Chem.*, 244, 6087, 1969.
29. **Hartz, J. W. and Deutsch, H. F.**, Preparation and physicochemical properties of human erythrocuprein, *J. Biol. Chem.*, 244, 4565, 1969.
30. **Briggs, R. G. and Fee, J. A.**, Further characterization of human erythrocyte superoxide dismutase, *Biochim. Biophys. Acta*, 537, 86, 1978.
31. **Briggs, R. G. and Fee, J. A.**, Sulfhydryl reactivity of human erythrocyte superoxide dismutase. On the origin of the unusual spectral properties of the protein when prepared by a procedure utilizing chloroform and ethanol for the precipitation of hemoglobin, *Biochim. Biophys. Acta*, 537, 100, 1978.
32. **Bannister, W. H., Anatasi, A., and Bannister, J. V.**, Human erythrocyte superoxide dismutase (erythrocuprein), in *Superoxide and Superoxide Dismutases*, Michelson, A., McCord, J. M., and Fridovich, I., Eds., Academic Press, London, 1977, 107.

32a. **Calabrese, L., Federici, G., Bannister, W. H., Bannister, J. V., Rotilio, G., and Finazzi-Agro, A.**, Labile sulfur in human superoxide dismutase, *Eur. J. Biochem.*, 56, 305, 1975.

33. **Marklund, S., Beckman, G., and Stigbrand, T.**, A comparison between the common type and a rare genetic variant of human cupro-zinc superoxide dismutase, *Eur. J. Biochem.*, 65, 415, 1976.
34. **Barikowiak, A., Lpyko, W., and Fried, R.**, A comparative characterization of cytosolic superoxide dismutase (SOD) from hog liver and erythrocytes, *Comp. Biochem. Physiol.*, 62B, 51, 1975.
35. **Abu-Erreish, G., Magnes, L., and Li, T. -K.**, Isolation and properties of superoxide dismutase from ram spermatozoa and erythrocytes, *Biol. Reprod.*, 18, 554, 1978.
36. **Crapo, J. D. and McCord, J. M.**, Oxygen-induced changes in pulmonary superoxide dismutase assayed by antibody titrations, *Am. J. Physiol.*, 231, 1196, 1976.
37. **Kanematsu, S. and Asada, K.**, Crystalline ferric superoxide dismutase from an anaerobic green sulfur bacterium, *Chlorobium thiosulfatophilum, FEBS Lett.*, 91, 94, 1978.
38. **Kanematsu, S. and Asada, K.**, Superoxide dismutase from an anaerobic photosynthetic bacterium, *Chromatium vinosum, Arch. Biochem. Biophys.*, 185, 473, 1978.
39. **Hatchikian, E. C. and Henry, Y. A.**, An iron-containing superoxide dismutase from the strict anaerobe *Desulfovibrio desulfuricans* (Norway 4), *Biochimie*, 59, 153, 1977.
40. **Lumsden, J., Cammack, R., and Hall, D. O.**, Purification and physiochemical properties of superoxide dismutase from two photosynthetic microorganisms, *Biochim. Biophys. Acta*, 438, 380, 1976.
41. **Anastasi, A., Bannister, J. V., and Bannister, W. H.**, Isolation and characterization of iron superoxide dismutase from *Bacillus megaterium, Int. J. Biochem.*, 7, 541, 1979.
42. **Kirby, T., Blum, J., Kahane, I., and Fridovich, I.**, Distinguishing between Mn-containing and Fe-containing superoxide dismutases in crude extracts of cells, *Arch. Biochem. Biophys.*, 201, 551, 1980.
43. **Brock, C. J., Harris, J. I., and Sato, S.**, Superoxide dismutase from *Bacillus stearothermophilus*. Preparation of stable apoprotein and reconstitution of fully active Mn enzyme, *J. Mol. Biol.*, 107, 175, 1976.

44. **Brock, C. J. and Walker, J. E.**, Superoxide dismutase from *Bacillus stearothermophilus.* Complete amino acid sequence of a manganese enzyme, *Biochemistry,* 19, 2873, 1980.
45. **Keele, B. B., Jr., McCord, J. M., and Fridovich, I.**, Superoxide dismutase from *Escherichia coli* B. A new manganese-containing enzyme, *J. Biol. Chem.,* 245, 6176, 1970.
46. **Pick, M., Rabani, J., Yost, F., and Fridovich, I.**, The catalytic mechanism of the manganese-containing superoxide dismutase of *Escherichia coli* studied by pulse radiolysis, *J. Am. Chem. Soc.,* 96, 7329, 1974.
47. **Fee, J. A., Shapiro, E. R., and Moss, T. H.**, Direct evidence for manganese (III) binding to the mangano superoxide dismutase of *Escherichia coli* B, *J. Biol. Chem.,* 251, 6157, 1976.
48. **Yost, F. J. and Fridovich, I.**, An iron-containing superoxide dismutase from *Escherichia coli., J. Biol. Chem.,* 248, 4905, 1973.
49. **Sylkhouse, T. O. and Fee, J. A.** Physical and chemical studies of bacterial superoxide dismutases. Purification and some anion binding properties of the iron-containing protein of *Escherichia coli* B, *J. Biol. Chem.,* 251, 5472, 1976.
50. **Dougherty, H. W., Sandowski, S. J., and Baker, E. E.**, A new iron-containing superoxide dismutase from *Escherichia coli, J. Biol. Chem.,* 253, 5220, 1978.
51. **Ichihara, K., Kusunose, E., Kusunose, M., and Mori, T.**, Superoxide dismutase from *Mycobacterium lepraemurium, J. Biochem. (Tokyo),* 81, 1427, 1977.
52. **Chikata, Y., Kusunose, E., Ichihara, K., and Kusunose, M.**, Purification of superoxide dismutases from *Mycobacterium phlei, Osaka City Med. J.,* 21, 127, 1975.
53. **Kusunose, M., Noda, Y., Ichihara, K., and Kusunose, E.**, Superoxide dismutase from *Mycobacterium* species, strain Takeo, *Arch. Microbiol.,* 108, 65, 1976.
54. **Kusunose, E., Ichihara, K., Noda, Y., and Kusunose, M.**, Superoxide dismutase from *Mycobacterium tuberculosis, J. Biochem. (Tokyo),* 80, 1343, 1976.
55. **Puget, K. and Michelson, A. M.**, Iron containing superoxide dismutases from luminous bacteria, *Biochimie,* 56, 1255, 1974.
56. **Asada, K., Yoshikawa, K., Takahashi, M., Maeda, Y., and Enmanji, K.**, Superoxide dismutases from a blue-green alga, *Plectonema boryanum, J. Biol. Chem.,* 250, 2801, 1975.
57. **Misra, H. P. and Keele, B. B., Jr.**, The purification and properties of superoxide dismutase from a blue-green alga, *Biochim. Biophys. Acta,* 379, 418, 1975.
58. **Yamakura, F.**, Purification, crystallization and properties of iron-containing superoxide dismutase from *Pseudomonas ovalis, Biochim. Biophys. Acta,* 422, 280, 1976.
59. **Lumsden, J. and Hall, D. O.**, Soluble and membrane-bound superoxide dismutases in a blue-green alga (*Spirulina*) and spinach, *Biochem. Biophys. Res. Commun.,* 58, 35, 1974.
60. **Britton, L., Malinowski, D. P., and Fridovich, I.**, Superoxide dismutase and oxygen metabolism in *Streptococcus faecalis* and comparisons with other organisms, *J. Bacteriol.,* 134, 229, 1978.
61. **Vance, P. G., Keele, B. B., Jr., and Rajagopalan, K. V.**, Superoxide dismutase from *Streptococcus mutans.* Isolation and characterization of two forms of the enzyme, *J. Biol. Chem.,* 247, 4782, 1972.
62. **Sato, S. and Harris, J. I.**, Superoxide dismutase from *Thermus aquaticus, Eur. J. Biochem.,* 73, 373, 1977.
63. **Sato, S. and Nakazawa, K.**, Purification and properties of superoxide dismutase from *Thermus thermophilus* HB8, *J. Biochem. (Tokyo),* 83, 1165, 1978.
64. **Misra, H. P. and Fridovich, I.**, Purification and properties of superoxide dismutase from a red alga, *Porphyridium cruentum, J. Biol. Chem.,* 252, 6421, 1977.
65. **Kanematsu, S. and Asada, K.**, Ferric and manganic superoxide dismutases in *Euglena gracilis, Arch. Biochem. Biophys.,* 195, 535, 1979.
66. **Lavelle, F., Durosay, P., and Michelson, A. M.**, Purification and properties of superoxide dismutase from the fungus *Pleurotus olearius, Biochimie,* 56, 451, 1974.
67. **Lavelle, F. and Michelson, A. M.**, Purification and study of the two superoxide dismutases from the fungus, *Pleurotus olearius, Biochimie,* 57, 375, 1975.
68. **Ravindranath, S. D. and Fridovich, I.**, Isolation and characterization of a manganese-containing superoxide dismutase from yeast, *J. Biol. Chem.,* 250, 6107, 1975.
69. **Salin, M. L. and Bridges, S. M.**, Isolation and characterization of an iron-containing superoxide dismutase form a eukaryote, *Brassica campestris, Arch. Biochem. Biophys.,* 201, 369, 1980.
70. **Sevilla, F., Gorge, J. L., and DelRio, L. A.**, in *Chemical and Biochemical Aspects of Superoxide and Superoxide Dismutase,* Bannister, J. V. and Hill, H. A. O., Eds., Elsevier/North-Holland, New York, 1980, 185.
71. **Salin, M. L., Day, E. D., Jr., and Crapo, J. D.**, Isolation and characterization of a manganese-containing superoxide dismutase from rat liver, *Arch. Biochem. Biophys.,* 187, 1978.
72. **Marklund, S.**, Purification and characterization of a manganese containing superoxide dismutase from bovine heart mitochondria, *Int. J. Biochem.,* 9, 299, 1978.

73. **McCord, J. M., Boyle, J. A., Day, E. D., Jr., Rizzolo, L. J., and Salin, M. L.**, A manganese-containing superoxide dismutase from human liver, in *Superoxide and Superoxide Dismutases*, Michelson, A. M., McCord, J. M., and Fridovich, I., Eds., Academic Press, London, 1977, 129.
74. **Crapo, J. D., McCord, J. M., and Fridovich, I.**, Preparation and assay of superoxide dismutases, *Meth. Enzymol.*, 53, 382, 1978.
75. **Fridovich, I.**, Superoxide dismutases, *Adv. Enzymol.*, 41, 35, 1974.
76. **Fridovich, I.**, Superoxide dismutases, *Annu. Rev. Biochem.*, 44, 147, 1975.
77. **Evans, H. J., Steinman, H. M. and Hill, R. L.**, Bovine erythrocyte superoxide dismutase. Isolation and characterization of tryptic, cyanogen bromide, and maleylated tryptic peptides, *J. Biol. Chem.*, 249, 7315, 1974.
78. **Steinman, H. M., Naik, V. R., Abernethy, J. L., and Hill, R. L.**, Bovine erythrocyte superoxide dismutase. Complete amino acid sequence, *J. Biol. Chem.*, 249, 7326, 1974.
79. **Abernethy, J. L., Steinman, H. M., and Hill, R. L.**, Bovine erythrocyte superoxide dismutase. Subunit structure and sequence location of the intrasubunit disulfide bond, *J. Biol. Chem.*, 249, 7339, 1974.
80. **Barra, D., Martini, F., Bannister, J. V., Schinina, M. E., Rotilio, G., Bannister, W. H., and Bossa, F.**, The complete amino acid sequence of human Cu/Zn superoxide dismutase, *FEBS Lett.*, 120, 53, 1980.
81. **Jabusch, J. R., Farb, D. L., Kerschensteiner, D. A., and Deutsch, H.**, Some sulfhydryl properties and primary structure of human erythrocyte superoxide dismutase, *Biochemistry*, 19, 2310, 1980.
82. **Johansen, J. T., Overballe-Petersen, C., Martin, B., Hassemann, V., and Svendsen, I.**, The complete amino acid sequence of copper, zinc, superoxide dismutase from *Saccharomyces cerevisiae, Carlsberg Res. Commun.*, 44, 201, 1979.
83. **Steinman, H. M.**, The amino acid sequence of copper-zinc superoxide dismutase from bakers' yeast, *J. Biol. Chem.*, 255, 6758, 1980.
84. **Ammer, D. and Lerch, K.**, Primary structure of horse liver superoxide dismutase, in *Chemical and Biochemical Aspects of Superoxide and Superoxide Dismutase*, Bannister, J. V. and Hill, H. A. O., Eds., Elsevier/North-Holland, New York, 1980, 230.
85. **Richardson, J. S., Thomas, K. A., Rubin, B. H., and Richardson, D. C.**, Crystal structure of bovine Cu, Zn superoxide dismutase at 3 Å resolution: chain tracing and metal ligands, *Proc. Natl. Acad. Sci. U.S.A.*, 72, 1349, 1975.
86. **Richardson, D. C.**, The three-dimensional structure of Cu, Zn superoxide dismutase, in *Superoxide and Superoxide Dismutases*, Michelson, A. M., McCord, J. M., and Fridovich, I., Eds., Academic Press, London, 1977, 217.
87. **Beem, K. M., Richardson, D. C., and Rajagopalan, K. V.**, Metal sites of copper-zinc superoxide dismutase, *Biochemistry*, 16, 1931, 1977.
88. **Hartz, J. W. and Deutsch, H. F.**, Subunit structure of human superoxide dismutase, *J. Biol. Chem.*, 247, 7043, 1972.
89. **Steinman, H. M.**, The amino acid sequence of mangano superoxide dismutase from *Escherichia coli* B, *J. Biol. Chem.*, 253, 8708, 1978.
90. **Tsuchihashi, M.**, Towards the understanding of blood catalase, *Biochem. Z.*, 140, 63, 1923.
91. **Stansell, M. J. and Deutsch, H. F.**, Preparaton of crystalline erythrocuprein and catalase from human erythrocytes, *J. Biol. Chem.*, 240, 4299, 1965.
92. **Stansell, M. J. and Deutsch, H. F.**, Physicochemical studies of crystalline human erythrocuprein, *J. Biol. Chem.*, 240, 4306, 1965.
93. **Markowitz, H., Cartwright, G. E., and Wintrobe, M. M.**, Studies on copper metabolism. XXVII. The isolation and properties of an erythrocyte cuproprotein (erythrocuprein), *J. Biol. Chem.*, 234, 40, 1959.
94. **Forman, H. J. and Fridovich, I.**, On the stability of bovine superoxide dismutase. The effects of metals, *J. Biol. Chem.*, 248, 2645, 1973.
95. **Malinowski, D. P. and Fridovich, I.**, Subunit association and side-chain reactivities of bovine erythrocyte superoxide dismutase in denaturing solvents, *Biochemistry*, 18, 5055, 1979.
96. **Klug, D., Rabani, J., and Fridovich, I.**, A direct demonstration of the catalytic action of superoxide dismutase through the use of pulse radiolysis, *J. Biol. Chem.*, 247, 4839, 1972.
97. **Fee, J. A. and Phillips, W. D.**, The behavior of holo- and apo-forms of bovine superoxide dismutase at low pH, *Biochim. Biophys. Acta*, 412, 26, 1975.
98. **Barra, D., Bossa, F., Marmocchi, F., Martini, F., Rigo, A., and Rotilio, G.**, Differential effects of urea on yeast and bovine copper, zinc superoxide dismutases, in relation to the extent of analogy of primary structures, *Biochem. Biophys. Res. Commun.*, 86, 1199, 1979.
99. **Lavelle, F., McAdam, M. E., Fielden, E. M., Roberts, P. B., Puget, K., and Michelson, A. M.**, A pulse-radiolysis study of the catalytic mechanism of the iron-containing superoxide dismutase from *Photobacterium leiognathi, Biochem. J.*, 161, 3, 1977.

100. **McAdam, M. E., Lavelle, F., Fox, R. A., and Fielden, E. M.,** A pulse-radiolysis study of the manganese-containing superoxide dismutase from *Bacillus stearothermophilus.* Further studies on the properties of the enzyme, *Biochem. J.,* 165, 81, 1977.
101. **Brock, C. J. and Harris, J. I.,** Superoxide dismutase from *Bacillus stearothermophilus*: reversible removal of manganese and its replacement by other metals, *Biochem. Soc. Trans.,* 5, 1537, 1977.
102. **Brock, C. J. and Walker, J. E.,** Superoxide dismutase from *Bacillus stearothermophilus*: metal binding and complete amino acid sequence, in *Chemical and Biochemical Aspects of Superoxide and Superoxide Dismutase,* Bannister, J. V. and Hill, H. A. O., Eds., Elsevier/North-Holland, New York, 1980, 237.
103. **Ose, D. E. and Fridovich, I.,** Superoxide dismutase. Reversible removal of manganese and its substitution by cobalt, nickel, or zinc, *J. Biol. Chem.,* 251, 1217, 1976.
104. **Ose, D. E. and Fridovich, I.,** Manganese-containing superoxide dismutase from *Escherichia coli*: reversible resolution and metal replacement, *Arch. Biochem. Biophys.,* 194, 360, 1979.
105. **Misra, H. P. and Fridovich, I.,** Inhibition of superoxide dismutases by azide, *Arch. Biochem. Biophys.,* 189, 317, 1978.
106. **Pile, J. and Dougherty, H. W.,** Ferrisuperoxide dismutase distribution in several bacteria, *J. Gen. Appl. Microbiol.,* 23, 303, 1977.
107. **McEuen, A. R., Hill, H. A. O., Dring, G. J., and Ingram, G. A.,** Comparison of methods for determining the prosthetic metal ion of bacterial superoxide dismutases in crude extracts, in *Chemical and Biochemical Aspects of Superoxide and Superoxide Dismutase,* Bannister, J. V. and Hill, H. A. O., Eds., Elsevier/North-Holland, New York, 1980, 272.
108. **Henry, L. E. A. and Hall, D. O.,** Superoxide dismutases in green algae: an evolutionary survey, *Plant and Cell Physiol.,* Special Issue 3, *Photosynthetic Organelles,* 377, 1977.
109. **Weisiger, R. A. and Fridovich, I.,** Mitochondrial superoxide dismutase. Site of synthesis and intramitochondrial localization, *J. Biol. Chem.,* 248, 4793, 1973.
110. **Hastings, J. W.,** Light to hide by: ventral luminescence to camouflage the silhouette, *Science,* 173, 1016, 1971.
111. **Drummond, M.,** Crown gall disease, *Nature (London),* 281, 343, 1979.
112. **Thomashow, M. F., Nutter, R., Postle, K., Chilton, M. D., Blattner, F. R., Powell, A., Gordon, M. P., and Nester, E. W.,** Recombination between higher plant DNA and Ti plasmid of *Agrobacterium tumefaciens, Proc. Natl. Acad. Sci. U.S.A.,* 77, 6448, 1980.
113. **Fridovich, I.,** personal communication, 1980.
114. **Bang, S. S., Woolkalis, M. J., and Baumann, P.,** Electrophoretic mobilities of superoxide dismutases from species of *Photobacterium, Beneckea, Vibrio,* and selected terrestrial enterobacteria, *Curr. Microbiol.,* 1, 371, 1978.
115. **McCord, J. M., Keele, B. B., Jr., and Fridovich, I.,** An enzyme-based theory of obligate anaerobiosis: the physiological function of superoxide dismutase, *Proc. Natl. Acad. Sci. U.S.A.,* 68, 1024, 1971.
116. **Winterbourn, C. C.,** Superoxide dismutase: potential for a wider role as an inhibitor of free radical reactions, in *Chemical and Biochemical Aspects of Superoxide and Superoxide Dismutase,* Bannister, J. V. and Hill, H. A. O., Eds., Elsevier/North-Holland, New York, 1980, 372.
117. **Hartley, B. S.,** Strategy and tactics in protein chemistry, *Biochem. J.,* 119, 805, 1970.
118. **Beale, G. and Knowles, J.,** *Extranuclear Genetics,* Edward Arnold, Ltd., London, 1978, 66.
119. **Smith, E. L.,** Evolution of enzymes, in *The Enzymes,* Vol. 1, 3rd ed., Boyer, P. D., Ed., Academic Press, New York, 1970, 267.
120. **Doolittle, R. F.,** Protein evolution, in *The Proteins,* Vol. 4, 3rd ed., Neurath, H. and Hill, R. L., Eds., Academic Press, New York, 1979, 1.
121. **Harris, J. I., Auffret, A. D., Northrup, F. D., and Walker, J. E.,** Structural comparisons of superoxide dismutase, *Europ. J. Biochem.,* 106, 297, 1980.
122. **Walker, J. E., Auffret, A. D., Brock, C. J., and Steinman, H. M.,** Structural comparisons of superoxide dismutases in *Chemical and Biochemical Aspects of Superoxide and Superoxide Dismutases,* Bannister, J. V. and Hill, H. A. O., Eds., Elsevier/North-Holland, New York, 1980, 212.
123. **Fridovich, I.,** Evidence for the symbiotic origin of mitochondria, *Life Sci.,* 14, 819, 1974.
124. **Uzzell, T. and Spolsky, C.,** Mitochondria and plasmids as endosymbionts: a revival of special creation? *Am. Sci.,* 62, 334, 1974.
125. **James, M. N. G.,** An X-ray crystallographic approach to enzyme structure and function, *Can. J. Biochem.,* 58, 251, 1980.
126. **Sherman, F. and Stewart, J. W.** The genetic control of yeast iso-1 and iso-2-cytochrome c after 15 years in *Biochemistry and Genetics of Yeast. Pure and Applied Aspects,* Bacila, M., Horecker, B. L., and Stoppani, A. O. M., Eds., Academic Press, New York, 1978, 273.
127. **Hase, T., Wakabayashi, S., Wada, K., and Matsubara, H.,** Amino acid sequence of *Aphanothece sacrum* ferredoxin II (minor component), *J. Biochem. (Tokyo),* 83, 761, 1978.

128. **Wakabayashi, S., Hase, T., Wada, K., Matsubara, H., Suzuki, K., and Takaichi, S.,** Amino acid sequences of two ferredoxins from poke weed, *Phytolacca americana, J. Biochem. (Tokyo),* 83, 1305, 1978.
129. **Kagamiyama, M., Sakakibara, R., Wada, M., Tanase, S., and Morino, Y.,** The complete amino acid sequence of mitochondrial aspartate aminotransferase from pig heart, *J. Biochem. (Tokyo),* 82, 291, 1977.
130. **Gehring, H., Christen, P., Eichele, G., Glor, M., Jansonius, J. N., Reimer, A. S., Smit, J. D. G., and Thaller, C.,** Isolation, crystallization and preliminary crystallographic data of aspartate aminotransferase from chicken heart mitochondria, *J. Mol. Biol.,* 115, 97, 1977.
131. **Barra, D., Bossa, F., Doonan, S., Fahmy, H. M. A., Hughes, G. J., Kakoz, K. Y., Martini, F., and Petruzelli, R.,** The structure of mitochondrial aspartate aminotransferase from pig heart and comparison with that of the cytoplasmic isozyme, *FEBS Lett.,* 83, 241, 1977.
132. **Fridovich, I.,** Chairman's introduction, in *Oxygen Free Radicals and Tissue Damage,* Ciba Found. Symp. 65. Excerpta Medica, Amsterdam, 1979, 1.
133. **Flohé, L.,** Glutathione peroxidase: fact and fiction, in *Oxygen Free Radicals and Tissue Damage,* CIBA Found. Symp. 65, Excerpta Medica, Amsterdam, 1979, 95.
134. **Tainer, J. A., Getzoff, E. D., Alden, C. J., Richardson, J. S., and Richardson, D. C.,** Geography of the active site of bovine Cu, Zn superoxide dismutase, *Fed. Proc.,* 39, (Abstr. 1946), 1976, 1980.
135. **Richardson, J. S.,** β-Sheet topology and the relatedness of proteins, *Nature (London),* 268, 495, 1977.
136. **Symonyan, M. A. and Nalbandyan, R. M.,** Some properties of erythrocuprein treated by organic solvents, *Biochim. Biophys. Acta,* 446, 432, 1976.
137. **Wood, E., Dalgleish, D., and Bannister, W.,** Bovine erythrocyte copper-zinc proteins. II. Physicochemical properties and circular dichroism, *Eur. J. Biochem.,* 18, 187, 1971.
138. **Rotilio, G., Calabrese, L., Bossa, F., Barra, D., Agro, A. F., and Mondovi, B.,** Properties of the apoprotein and role of copper and zinc in protein conformation and enzyme activity of bovine superoxide dismutase, *Biochemistry,* 11, 2182, 1972.
139. **Cass, A. E. G., Hill, H. A. O., Bannister, J. V., and Bannister, W. H.,** Zinc (II) binding to apo-(bovine erythrocyte superoxide dismutase), *Biochem. J.,* 177, 477, 1979.
140. **Bray, R. C., Cockle, S. A., Fielden, E. M., Roberts, P. B., Rotilio, G., and Calabrese, L.,** Reduction and inactivation of superoxide dismutase by hydrogen peroxide, *Biochem. J.,* 139, 43, 1974.
140a. **Forman, H. J., Evans, H. J., Hill, R. L., and Fridovich, I.,** Histidine at the active site of superoxide dismutase, *Biochemistry,* 12, 823, 1973.
141. **Blumberg, W. E., Peisach, J., Eisenberger, P. and Fee, J. A.,** Superoxide dismutase, a study of the electronic properties of the copper and zinc by X-ray absorption spectroscopy, *Biochemistry,* 17, 1842, 1978.
142. **Hodgson, E. K. and Fridovich, I.,** The interaction of bovine erythrocyte superoxide dismutase with hydrogen peroxide: inactivation of the enzyme, *Biochemistry,* 14, 5294, 1975.
143. **Lynch, R. E. and Fridovich, I.,** Autoinactivation of xanthine oxidase. The role of superoxide radical and hydrogen peroxide, *Biochim. Biophys. Acta,* 571, 195, 1979.
144. **Malinowski, D. P. and Fridovich, I.,** Chemical modification of arginine at the active site of the bovine erythrocyte superoxide dismutase, *Biochemistry,* 18, 5909, 1979.
145. **Malinowski, D. P. and Fridovich, I.,** An essential arginine residue at the active site of copper-zinc superoxide dismutases, in *Chemical and Biochemical Aspects of Superoxide and Superoxide Dismutase,* Bannister, J. V. and Hill, H. A. O., Eds., Elsevier/North-Holland, New York, 1980, 299.
146. **Borders, C. L., Jr. and Johansen, J. T.,** Essential arginyl residues in Cu, Zn superoxide dismutase from *Saccharomyces cerevisiae, Carlsberg Res. Commun.,* 45, 185, 1980.
147. **Borders, C. L., Jr. and Johansen, J. T.,** Identification of Arg-143 as the essential arginyl residue in yeast Cu,Zn superoxide dismutase by use fo a crhomophoric arginine reagent, *Biochem. Biophys. Res. Commun.,* 96, 1071, 1980.
148. **Agro, A. F., Cocco, D., Calabrese, L., Bannister, W. H., and Bossa, F.,** Aromatic sidechains in Cu-Zn superoxide dismutases, *Int. J. Biochem.,* 8, 389, 1977.
149. **Dickerson, R. E. and Timkovich, R.,** Cytochromes *c* in *The Enzymes,* Vol. XI (Part A), 3rd Ed., Boyer, P. D., Ed., Academic Press, New York, 1975, 397.
150. **Morita, Y. and Asada, K.,** Preliminary crystallographic data for spinach superoxide dismutase, *J. Mol. Biol.,* 86, 685, 1974.
151. **Bossa, F., Barra, D., Marmocchi, F., Martini, F., Rotilio, G., Bannister, F. V., and Bannister, W. H.,** Studies on the protein structure of copper-zinc superoxide dismutases, in *Chemical and Biochemical Aspects of Superoxide and Superoxide Dismutase,* Bannister, J. V. and Hill, H. A. O., Eds., Elsevier/North-Holland, New York, 1980, 223.
152. **Stryer, L.,** Fluorescence energy transfer as a spectroscopic ruler, *Ann. Rev. Biochem.,* 47, 819, 1978.

153. **Huber, W. and Saifer, M. G. P.**, Orgotein, the drug version of bovine Cu-Zn superoxide dismutase. I. a summary account of safety and pharmacology in laboratory animals, in *Superoxide and Superoxide Dismutases,* Michelson, A. M., McCord, J. M., and Fridovich, I., Eds., Academic Press, New York, 1977, 517.
154. **Kahn, P. M.**, Enzyme electrophoresis on cellulose acetate gel: zymogram patterns in man-mouse and man-chinese hamster somatic cell hybrids, *Arch. Biochem. Biophys.,* 145, 470, 1971.
155. **Marmocchi, F., Venardi, G., Bossa, F., Rigo, A., and Rotilio, G.**, Dissociation of Cu-Zn superoxide dismutase into monomers by urea. Evidence from gel filtration and molecular hybridization, *FEBS Lett.,* 94, 109, 1978.
156. **Malinowski, D. P. and Fridovich, I.**, Bovine erythrocyte superoxide dismutase: diazo coupling, subunit interactions, and electrophoretic variants, *Biochemistry,* 18, 237, 1979.
157. **Tegelström, H.**, Interspecific hybridization in vitro of superoxide dismutase from various species, *Hereditas,* 81, 185, 1975.
158. **Suzuki, K. and Harris, J. I.**, Hybridization of glyceraldehyde-3-phosphate dehydrogenase, *J. Biochem. (Tokyo),* 77, 587, 1975.
159. **Asada, K., Kanematsu, S., and Uchida, K.**, Superoxide dismutases in photosynthetic organisms: absence of the cuprozinc enzyme in eukaryotic algae, *Arch. Biochem. Biophys.,* 179, 243, 1977.
160. **Baret, A., Michel, P., Imbert, M. R., Morcellet, J. L., and Michelson, A. M.**, A radioimmunoassay for copper containing superoxide dismutase, *Biochem. Biophys. Res. Commun.,* 88, 337, 1979.
161. **Behar, D., Czapski, G., Rabani, J., Dorfman, L. M., and Schwarz, H. A.**, The acid dissociation constant and decay kinetics of the perhydroxyl radical, *J. Phys. Chem.,* 74, 3209, 1970.
162. **Rotilio, G., Bray, R. C., and Fielden, E. M.**, A pulse radiolysis study of superoxide dismutase, *Bichim. Biophys. Acta,* 268, 605, 1972.
163. **Fielden, E. M., Roberts, P. B., Bray, R. C., Lowe, D. J., Mautner, G. N., Rotilio, G., and Calabrese, L.**, The mechanism of action of superoxide dismutase from pulse radiolysis and electron paramagnetic resonance, *Biochem. J.,* 139, 49, 1974.
164. **Bannister, J. V., Bannister, W. H., Bray, R. C., Fielden, E. M., Roberts, P. B., and Rotilio, G.**, The superoxide dismutase activity of human erythrocuprein, *FEBS Lett.,* 32, 303, 1973.
165. **Rigo, A., Viglino, P., and Rotilio, G.**, Kinetic study of O_2^- dismutation by bovine superoxide dismutase. Evidence for saturation of the catalytic site by O_2^-, *Biochem. Biophys. Res. Commun.,* 63, 1013, 1975.
166. **Plonka, A., Metodiewa, D., and Jeziorowska, A.**, ESR evidence of paramagnetic intermediate formation from water radiolysis products and SOD, *Biochem. Biophys. Res. Commun.,* 81, 1344, 1978.
167. **Plonka, A., Metodiewa, D., and Gasyna, Z.**, ESR studies on oxidation state changes of copper in superoxide dismutase during reactions with water radiolysis products at cryogenic temperatures, *Biochim. Biophys. Acta,* 612, 299, 1980.
168. **Fee, J. A.**, Structure-function relationships in superoxide dismutases, in *Superoxide and Superoxide Dismutases,* Michelson, A. M., McCord, J. M., and Fridovich, I., Eds., Academic Press, London, 1977, 173.
169. **Fee, J. A. and Gaber, B. P.**, Anion binding to bovine erythrocyte superoxide dismutase, *J. Biol. Chem.,* 247, 60, 1972.
170. **Gaber, B. P., Brown, R. D., Koenig, S. H., and Fee, J. A.**, Nuclear magnetic relaxation dispersion in protein solutions. V. Bovne erythrocyte superoxide dismutase, *Biochim. Biophys. Acta,* 271, 1, 1972.
171. **Fee, J. A. and Ward, R. L.**, Evidence for a coordination position available to solute molecules on one of the metals at the active center of reduced bovine superoxide dismutase, *Biochem. Biophys. Res. Commun.,* 71, 427, 1976.
172. **Cockle, S. A. and Bray, R. C.**, Do all the copper atoms in bovine superoxide dismutase function in catalysis? in *Superoxide and Superoxide Dismutases,* Michelson, A. M., McCord, J. M., and Fridovich, I., Eds., Academic Press, London, 1977, 215.
173. **Rigo, A., Viglino, P., Calabrese, L., Cocco, D., and Rotilio, G.**, The binding of copper ions to copper-free bovine superoxide dismutase. Copper distribution in protein samples recombined with less than stoichiometric copper ion/protein ratio, *Biochem. J.,* 161, 27, 1977.
174. **Rigo, A., Terenzi, M., Viglino, P., Calabrese, L., Cocco, D., and Rotilio, G.**, The binding of copper ions to copper-free bovine superoxide dismutase. Properties of the protein recombined with increasing amounts of copper ions, *Biochem. J.,* 161, 31, 1977.
175. **Rigo, A., Viglino, P., Bonori, M., Cocco, D., Calabrese, L., and Rotilio, G.**, The binding of copper ions to copper-free bovine superoxide dismutase. Kinetic aspects, *Biochem. J.,* 169, 277, 1978.
176. **Zukin, R. S., Hartig, P. R., and Koshland, D. E., Jr.**, Use of a distant reporter group as evidence for a conformational change in a sensory receptor, *Proc. Natl. Acad. Sci. U.S.A.,* 74, 1932, 1977.
177. **Bannister, J. V., Anastasi, A., and Bannister, W. H.**, Active subunits from superoxide dismutase, *Biochem. Biophys. Res. Commun.,* 18, 469, 1978.

178. **Guidotti, G., Konigsberg, W., and Craig, L. C.,** On the dissociation of normal adult human hemoglobin, *Proc. Natl. Acad. Sci. U.S.A.,* 150, 774, 1963.
179. **Benesch, R., Benesch, R. E., and Tyuma, I.,** Subunit exchange and ligand binding. II. The mechanism of the allosteric effect in hemoglobin, *Proc. Natl. Acad. Sci. U.S.A.,* 56, 1268, 1966.
180. **Marmocchi, F., Venardi, G., Caulini, G., and Rotilio, G.,** Enzyme activity of superoxide dismutase protomers, *FEBS Lett.,* 44, 337, 1974.
181. **Rigo, A., Marmocchi, F., Cocco, D., Viglino, P., and Rotilio, G.,** On the quaternary structure of copper-zinc superoxide dismutases. Reversible dissociation into protomers of the isozyme I from wheat germ, *Biochemistry,* 17, 534, 1978.
182. **Beauchamp, C. O. and Fridovich, I.,** Superoxide dismutase: improved assays and an assay applicable to acrylamide gels, *Anal. Biochem.,* 44, 276, 1971.
183. **Beem, K. M., Rich, W. E., and Rajagopalan, K. V.,** Total reconstitution of copper-zinc superoxide dismutase, *J. Biol. Chem.,* 249, 7298, 1974.
184. **Fee, J. A.,** Studies on the reconstitution of bovine erythrocyte superoxide dismutase. IV. Preparation and some properties of the enzyme in which Co (II) is substituted for Zn (II), *J. Biol. Chem.,* 248, 4229, 1973.
185. **Lippard, S. J., Burger, A. R., Ugurbil, K., Pantoliano M. W., and Valentine, J. S.,** Nuclear magnetic resonance and chemical modification studies of bovine erythrocyte superoxide dismutase: evidence for zinc-promoted organization of the active site structure, *Biochemistry,* 16, 1136, 1977.
186. **Valentine, J. S. and Pantoliano, M. W.,** Metal binding properties of copper-zinc superoxide dismutase, in *Chemical and Biochemical Aspects of Superoxide and Superoxide Dismutase,* Bannister, J. V. and Hill, H. A. O., Eds., Elsevier/North-Holland, New York, 1980, 254.
187. **Cass, A. E. G., Hill, H. A. O., Bannister, J. V., Bannister, W. H., Hasemann, V., and Johansen, J. T.,** The exchange of histidine C-2 protons in superoxide dismutases. A novel method for assigning histidine-metal ligands in proteins, *Biochem. J.,* 183, 127, 1979.
188. **Haschemeyer, R. H. and Haschemeyer, A. E. V.,** *Proteins, a Guide to Study by Physical and Chemical Methods,* John Wiley & Sons, New York, 1973, 217 and 301.
189. **Fee, J. A.,** Studies on the reconstitution of bovine erythrocyte superoxide dismutase. III. Evidence for a strong interdependence between Cu^{2+} and Zn^{2+} binding in the expression of the spectroscopic properties of the native protein and for a close proximity of the Zn^{2+} and Cu^{2+} sites, *Biochim. Biophys. Acta,* 295, 107, 1973.
190. **Rotilio, G., Finazzi-Agro, A., Calabrese, L., Bossa, F., Guerrieri, P., and Mondovi, B.,** Studies of the metal sites of copper proteins. Ligands of copper in hemocuprein, *Biochemistry,* 10, 616, 1971.
191. **Stokes, A. M., Hill, H. A. O., Bannister, W. H., and Bannister, J. V.,** Nuclear magnetic resonance spectra of human and bovine superoxide dismutases, *FEBS Lett.,* 32, 119, 1973.
192. **Fee, J. A. and Briggs, R. G.,** Studies on the reconstitution of bovine erythrocyte superoxide dismutase. V. Preparation and properties of derivatives in which both zinc and copper sites contain copper, *Biochim. Biophys. Acta,* 400, 439, 1975.
193. **Rotilio, G., Calabrese, L., Mondovi, B., and Blumberg, W. E.,** Electron paramagnetic resonance studies of cobalt-copper bovine superoxide dismutase, *J. Biol. Chem.,* 249, 3157, 1974.
194. **Calabrese, L., Cocco, D., Morpurgo, L., Mondovi, B., and Rotilio, G.,** Cobalt bovine superoxide dismutase. Reactivity of the cobalt chromophore in the copper-containing and in the copper-free enzyme, *Eur. J. Biochem.,* 64, 465, 1976.
195. **Cass, A. E. G., Hill, H. A. O., Smith, B. E., Bannister, J. V., and Bannister, W. H.,** Investigation of the structure of bovine erythrocyte superoxide dismutase by ^1H nuclear magnetic resonance spectroscopy, *Biochemistry,* 16, 3061, 1977.
196. **Bannister, J. V., Bannister, W. H., Cass, A. E. G., Hill, H. A. O., and Johansen, J. T.,** A comparison of the active site structures of bovine, human, swordfish and yeast copper-zinc superoxide dismutases, in *Chemical and Biochemical Aspects of Superoxide and Superoxide Dismutases,* Bannister, J. V. and Hill, H. A. O., Eds., Elsevier/North-Holland, New York, 1980, 284.
197. **Carrico, R. J. and Deutsch, H. F.,** The presence of zinc in human cytocuprein and some properties of the apoprotein, *J. Biol. Chem.,* 245, 723, 1970.
198. **Marmocchi, F., Caulini, G., Venadi, G., Cocco, D., Calabrese, L., and Rotilio, G.,** The effect of the presence of the metal prosthetic groups on the subunit structure of bovine superoxide dismutase in sodium docecyl sulfate, *Physiol. Chem. Phys.,* 7, 465, 1975.
199. **Roberts, G. C. K. and Jardetsky, O.,** Nuclear magnetic resonance spectroscopy of amino acids, peptides, and proteins, *Adv. Protein Chem.,* 25, 447, 1970.
200. **Burger, A. R., Lippard, S. J., Pantoliano, M. W., and Valentine, J. S.,** Nuclear magnetic resonance study of the exchangeable histidine protons in bovine and wheat germ superoxide dismutases, *Biochemistry,* 19, 4139, 1980.
201. **Valentine, J. S., Pantoliano, M. W., McDonnell, P. J., Burger, A. R., and Lippard, S. J.,** pH-dependent migration of copper (II) to the vacant zinc-binding site of zinc-free bovine erythrocyte superoxide dismutase, *Proc. Natl. Acad. Sci. U.S.A.,* 76, 4245, 1979.

202. **Calabrese, L., Cocco, D., and Desideri, A.,** A novel Co (II) binding site in copper-free superoxide dismutase. Evidence for binding of cobalt at the copper binding site, *FEBS Lett.*, 106, 142, 1979.
203. **Moss, T. H. and Fee, J. A.,** On the magnetic properties of cobalt substituted bovine superoxide dismutase derivatives, *Biochem. Biophys. Res. Commun.*, 66, 799, 1975.
204. **Desideri, A., Cerdonio, M., Magno, F., Vitale, S., Calabrese, L., Cocco, D., and Rotilio, G.,** A magnetic susceptibility of Cu (II)-Co (II) superoxide dismutase, *FEBS Lett.*, 83, 1978.
204a. **Fee, J. A., Peisach, J., and Mims, W. B.,** Superoxide dismutase: examination of the metal binding sites by electron spin echo spectroscopy, *J. Biol. Chem.*, 256, 1910, 1981.
205. **Bailey, D. B., Ellis, P. D., and Fee, J. A.,** Cadmium-113 nuclear magnetic resonance studies of cadmium-substituted derivatives of bovine superoxide dismutase, *Biochemistry*, 19, 591, 1980.
206. **McAdam, M. E., Fielden, E. M., Lavelle, F., Calabrese, L., Cocco, D., and Rotilio, G.,** The involvement of the bridging imidazolate in the catalytic mechanism of action of bovine superoxide dismutase, *Biochem. J.*, 167, 271, 1977.
207. **Calabrese, L., Cocco, D., Morpurgo, L., Mondovi, B., and Rotilio, G.,** Reversible uncoupling of the copper and cobalt spin systems in cobalt bovine superoxide dismutase at low pH, *FEBS Lett.*, 59, 29, 1975.
208. **Pantoliano, M. W., McDonnell, P. J., and Valentine, J. S.,** Reversible loss of metal ions from the zinc binding site of copper-zinc superoxide dismutase. The low pH transition, *J. Am. Chem. Soc.*, 101, 6454, 1979.
209. **Bauer, R., Demeter, I., Hasemann, V., and Johansen J. T.,** Structural properties of the zinc site in Cu,Zn-superoxide dismutase; perturbed angular correlation of gamma ray spectroscopy on the Cu, ^{111}Cd-superoxide dismutase derivative, *Biochem. Biophys. Res. Commun.*, 94, 1296, 1980.
210. **Hill, H. A. O., Lee, W. -K., Bannister, J. V., and Bannister, W. H.,** Investigation of human erythrocyte superoxide dismutase by ^{1}H nuclear-magnetic-resonance spectroscopy, *Biochem. J.*, 185, 245, 1980.
211. **Cass, A. E. G., Hill, H. A. O., Smith, B. E., Bannister, J. V., and Bannister, W. H.,** Carbon-2 proton exchange at histidine-41 in bovine erythrocyte superoxide dismutase, *Biochem. J.*, 165, 587, 1977.
212. **Stoesz, J. D., Malinowski, D. P., and Redfield, A. G.,** Nuclear magnetic resonance study of solvent exchange and nuclear Overhauser effect of the histidine protons of bovine superoxide dismutase, *Biochemistry*, 18, 4669, 1979.
213. **Lönnerdal, B., Keen, C. L., and Hurley, L. S.,** Isoelectric focusing of superoxide dismutase isozymes, *FEBS Lett.*, 108, 51, 1979.
214. **Beckman, G.,** Population studies in northern Sweden. VI. Polymorphism of superoxide dismutase, *Hereditas*, 73, 305, 1973.
215. **Beckman, G., Beckman, L., and Nilsson, L. -O.,** A rare homozygous phenotype of superoxide dismutase, SOD 2, *Hereditas*, 75, 138, 1973.
216. **Beckman, G. and Pakarinen, A.,** Superoxide dismutase, a population study, *Hum. Hered.*, 23, 346, 1973.
217. **Tan, Y. H., Tischfield, J., and Ruddle, F. H.,** The linkage of genes for the human interferon-induced antiviral protein and indophenol oxidase-B traits to chromosome G-21, *J. Exp. Med.*, 137, 317, 1973.
218. **Creagan, R., Tischfield, J., Ricciuti, F., and Ruddle, F. H.,** Chromosome assignments of genes in man using mouse-human somatic cell hybrids: mitochondrial superoxide dismutase (indophenol oxidase-B, tetrameric) to chromosome 6, *Humangenetik*, 20, 203, 1973.
219. **Sinet, P. -M.,** SOD genes in humans: chromosome localization and electrophoretic variants, in *Superoxide and Superoxide Dismutases*, Michelson, A. M., McCord, J. M., and Fridovich, I., Eds., Academic Press, London, 1977, 459.
220. **Beem, K. M., Richardson, J. S., and Richardson, D. C.,** Manganese superoxide dismutases from *Escherichia coli* and from yeast mitochondria: preliminary X-ray crystallographic studies, *J. Mol. Biol.*, 105, 327, 1976.
221. **Bridgen, J., Harris, J. I., and Kolb, E.,** Superoxide dismutase from *Bacillus stearothermophilus*: crystallization and preliminary X-ray diffraction studies, *J. Mol. Biol.*, 105, 333, 1976.
222. **Smit, J. D. G., Pulver-Sladek, J., and Jansonius, J. N.,** Preliminary X-ray diffraction studies of orthorhombic crystals of superoxide dismutase from *Bacillus stearothermophilus*, *J. Mol. Biol.*, 112, 491, 1977.
223. **Yamakura, F., Suzuki, K., and Mitsui, Y.,** Crystallization and preliminary crystal data of iron-containing superoxide dismutase from *Pseudomonas ovalis*, *J. Biol. Chem.*, 251, 4792, 1976.
224. **Yamakura, F., Suzuki, K., Petsko, G. A., and Tsernoglou, D.,** Metal replacement studies and crystallographic data on iron-superoxide dismutase from *Pseudomonas ovalis*, in *Chemical and Biochemical Aspects of Superoxide and Superoxide Dismutase*, Bannister, J. V. and Hill, H. A. O., Eds., Elsevier/North-Holland, New York, 1980, 242.

225. **Powers, T. B., Slykhouse, T. O., Fee, J. A., and Ludwig, M. L.**, Characterization of an orthorhombic crystal form of iron-containing superoxide dismutase from *Escherichia coli* B, *J. Mol. Biol.*, 123, 689, 1978.
226. **Ambler, R. P., Daniel, M., Hermoso, J., Meyer, T. E., Bartsch, R. F., and Kamen, M. D.**, Cytochrome c_2 sequence variation among the recognized species of purple nonsulphur photosynthetic bacteria, *Nature (London)*, 278, 659, 1979.
227. **Steinman, H. M. and Hill, R. L.**, Sequence homologies among bacterial and mitochondrial superoxide dismutases, *Proc. Natl. Acad. Sci. U.S.A.*, 70, 3725, 1973.
228. **Bridgen, J., Harris, J. I., and Northrop, F.**, Evolutionary relationships in superoxide dismutases, *FEBS Lett.*, 49, 393, 1975.
229. **Bruschi, M., Hatchikian, E. C., Bonicel, J., Bovier-Lapierre, G., and Couchoud, P.**, The N-terminal sequence of superoxide dismutase from the strict anaerobe *Desulfovibrio desulfuricans*, *FEBS Lett.*, 76, 121, 1977.
230. **McLachlan, A. D.**, Quantum chemistry and protein folding: the art of the possible, *Int. J. Quantum Chem.*, 12 (Suppl. 1), 371, 1977.
231. **Beckman, G. and Holn, S.**, Immunological differences between human superoxide dismutase isozymes, *Hereditas*, 80, 1, 1975.
232. **Bang, S. S., Baumann, P., and Nealson, K. H.**, Phenotypic characterization of *Photobacterium logei* (sp. nov.), a species related to *P. fischeri*, *Curr. Microbiol.*, 1, 285, 1978.
233. **Britton, L. and Fridovich, I.**, Streptomycin: irreversible association with superoxide dismutases, *Arch. Biochem. Biophys.*, 191, 198, 1978.
234. **Yamakura, F. and Suzuki, K.**, Reconstitution of iron-superoxide dismutase, *Biochem. Biophys. Res. Commun.*, 72, 1108, 1976.
235. **Yamakura, F.**, A study on the reconstitution of iron-superoxide dismutase from *Pseudomonas ovalis*, *J. Biochem. (Tokyo)*, 83, 849, 1978.
236. **Yamakura, F. and Suzuki, K.**, Cadmium, chromium, and manganese replacement for iron in iron-superoxide dismutase from *Pseudomonas ovalis*, *J. Biochem. (Tokyo)*, 88, 191, 1980.
237. **Puget, K., Lavelle, F., and Michelson, A. M.**, Superoxide dismutases from procaryote and eucaryote bioluminescent organisms, in *Superoxide and Superoxide Dismutases*, Michelson, A. M., McCord, J. M., and Fridovich, I., Eds., Academic Press, New York, 1977, 139.
238. **Murphy, J. F. and Kies, M. W.**, Note on spectrophotometric determination of proteins in dilute solutions, *Biochim. Biophys. Acta*, 45, 382, 1960.
239. **Vanopdenbosch, B. and Crichton, R. R.**, Chemical modification studies on a ferri-superoxide dismutase from marine bacteria, in *Superoxide and Superoxide Dismutases*, Michelson, A. M., McCord, J. M., and Fridovich, I., Eds., Academic Press, London, 1977, 199.
240. **Vanopdenbosch, B. and Crichton, R. R.**, Chemical modification studies on a ferri-superoxide dismutase from marine bacteria, *FEBS Lett.*, 82, 42, 1977.
241. **McAdam, M. E., Fox, R. A., Lavelle, F., and Fielden, E. M.**, A pulse-radiolysis study of the manganese-containing superoxide dismutase from *Bacillus stearothermophilus*. A kinetic model for the enzyme action, *Biochem. J.*, 165, 71, 1977.
242. **Keele, B. B., Jr., Giovagnoli, C., and Rotilio, G.**, CD spectra and redox reactions of superoxide dismutase from *Escherichia coli* B: evidence for a Mn (III) enzyme, *Physiol. Chem. Phys.*, 7, 1, 1975.
243. **Villafranca, J. J., Yost, F. J., Jr., and Fridovich, I.**, Magnetic resonance studies of manganese (III) and iron (III) superoxide dismutases. Temperature and frequency dependence of proton relaxation rates of water, *J. Biol. Chem.*, 249, 3532, 1974.
244. **Villafranca, J. J.**, ESR spectra of Fe (III)-superoxide dismutase with special reference to the electron spin relaxation time of Fe (III), *FEBS Lett.*, 62, 230, 1976.

Chapter 3

MEASURING THE ACTIVITY OF SUPEROXIDE DISMUTASES: AN EMBARRASSMENT OF RICHES

Irwin Fridovich

TABLE OF CONTENTS

I. INTRODUCTION

Superoxide dismutases are unique in that their substrate is an unstable free radical. This dictates the need for devious strategies in assaying its activity. One cannot, as with most enzymes, prepare a stable stock solution of the substrate and then observe the acceleration of its conversion to product by the enzyme. Given this challenge to human ingenuity, it is not surprising that not one, but many means of circumventing the instability of the substrate radical have been devised. The multiplicity of available assays reflects not only the questing human spirit but also the truth that none of the available assays is entirely satisfactory, so that the stimulating need for improvements has always been present. We will, in summarizing the known assays, cluster them according to the general principles upon which they are based and point out the potential drawbacks, as well as the advantages, of each. In virtually all cases, once the principles of the assay and of the potential interferences are well understood, it is relatively easy to avoid or at least to control for them. A review of some of these methods was published in 1978.[1]

II. XANTHINE OXIDASE — CYTOCHROME *c* ASSAY

One way to largely circumvent the instability of $O_2^{\overline{\cdot}}$ is to continuously generate it in the assay solution. If that solution also contains an indicating scavenger for $O_2^{\overline{\cdot}}$, the rate of modification of that scavenger will be a measure of the rate of production of $O_2^{\overline{\cdot}}$. If the concentration of the indicating scavenger is varied, we note that its rate of modification increases to a limit. This is due to a competition between the spontaneous dismutation of $O_2^{\overline{\cdot}}$, to $O_2^{\overline{\cdot}}$ + H_2O_2 and the reaction with the indicating scavenger. As the concentration of the scavenger is raised, it traps an ever-increasing proportion of the $O_2^{\overline{\cdot}}$ made, while a smaller fraction dismutes. The limit is reached when essentially all of the $O_2^{\overline{\cdot}}$ reacts with the indicating scavenger and none is lost to the dismutation reaction.

Since the spontaneous dismutation is markedly pH dependent,[2] the partition between the spontaneous dismutation and the indicating scavenger will also be pH dependent. Furthermore, since the spontaneous dismutation is second order in the oxygen radical, whereas its reaction with the indicating scavenger is first order with respect to the radical, the partition between dismutation and scavenging will depend upon the steady-state concentration of the radical. Higher steady-state concentrations of $O_2^{\overline{\cdot}}$ will favor the dismutation reaction and higher concentrations of the indicating scavenger will then be needed to compete successfully against the dismutation. These points are considered in order to emphasize the importance of such variables as pH, rate of production of $O_2^{\overline{\cdot}}$, and the concentration of the indicating scavenger, on assays for SOD.

Cytochrome *c* is readily reduced by $O_2^{\overline{\cdot}}$ and serves as a useful indicating scavenger.[3-5] SOD then acts to inhibit cytochrome *c* reduction, because SOD catalyzes the dismutation and so competes with the cytochrome *c* for the available $O_2^{\overline{\cdot}}$. In practice, one unit of SOD is defined as that amount which causes 50% inhibition of cytochrome *c* reduction, under carefully specified conditions. The standard assay method used 10 μM ferricytochrome *c*, 50 μM xanthine, 100 μM EDTA, 50 mM potassium phosphate, and sufficient milk xanthine oxidase (∼5 nM) to give a rate of increase in absorbance of 0.025/min at pH 7.8 and at 25°C.[5]

Native milk xanthine oxidase can carry out both univalent and divalent reductions of O_2. The proportion of the electron flow through the enzyme, which goes to the univalent pathway, and hence to $O_2^{\overline{\cdot}}$ production, depends upon pH, pO_2, and to a lesser extent upon substrate concentration. Raising the pH and the pO_2 increases the

$O_2^{\cdot-}$ production markedly while raising the substrate concentration decreases it somewhat.[6] At pH 7.8, in solutions equilibrated with air, approximately one sixth of the electrons traversing xanthine oxidase go the univalent route.

Xanthine oxidase is susceptible to an activity-dependent inactivation by heavy metals, which are generally present as contaminants of phosphate buffers.[7] For this reason EDTA or some other chelating agent is usually incorporated in the assay mixtures. Milk and incompletely purified xanthine oxidase contain lactoperoxidase, which can catalyze the peroxidation of ferrocytochrome *c* and thus interfere with the xanthine oxidase-cytochrome *c* assay of SOD. Xanthine oxidase purified according to Waud et al.[8] has been found free of lactoperoxidase. In the event of a lactoperoxidase impurity in the xanthine oxidase being used, catalase can be added to eliminate its effect on the rate of accumulation of ferrocytochrome *c*. Of course, the catalase so used must be free of SOD. Some commercial preparations of catalase have been found wanting in this respect.[9]

There is nothing specific about the reduction of cytochrome *c* by $O_2^{\cdot-}$. Many common reductants are capable of reducing cytochrome *c* and quite a few enzymes can reduce it directly. It is crucial, for the SOD assay under discussion, that native xanthine oxidase does not directly reduce cytochrome *c*. This has the consequence that all of the cytochrome *c* reduction caused by xanthine oxidase is oxygen dependent, is mediated by $O_2^{\cdot-}$, and is inhibitable by SOD. In contrast, deflavoxanthine oxidase can directly reduce cytochrome *c*. Thus, to the extent that xanthine oxidase has been converted to the deflavo derivative, one will see the appearance of SOD-resistant cytochrome *c* reduction. Preparations which have been stored for long periods, frozen and thawed repeatedly or exposed to high salt concentrations, may be found contaminated with the deflavo derivative.

There is yet another way, aside from deflavination of the enzyme, to circumvent the $O_2^{\cdot-}$-dependency of cytochrome *c* reduction by xanthine oxidase and that is to add electron-carrying compounds, such as menadione, methylene blue, or flavin mononucleotide (FMN). These compounds can be reduced by xanthine oxidase, much as can O_2, yielding reduced forms which can, in turn, reduce cytochrome *c*. Such compounds will therefore impart to xanthine oxidase an O_2-independent and SOD-resistant route of cytochrome *c* reduction. Since crude extracts of cells or tissues may contain flavins or quinones, able to act in this way, they may decrease the sensitivity of the assay to SOD. Dialysis of extracts prior to assay will eliminate this problem.

There are other potential, but avoidable difficulties. Thus anything which inhibits xanthine oxidase will decrease the rate of cytochrome *c* reduction and thus give the appearance of SOD-like activity. It s simple to control for this effect by observing the effect of test samples on the rate of the xanthine oxidase reaction, followed in terms of urate production at 295 nm. Further, anything capable of directly reducing cytochrome *c* will impart an SOD-insensitive baseline rate. Dialysis of samples will remove low-molecular-weight reductants. Cytochrome *c* oxidase or peroxidase activity in the test samples will also interfere. Since these are heme enzymes strongly inhibited by cyanide, their effect can be eliminated by levels of cyanide ($\sim 5 \times 10^{-5} M$) too low to significantly inhibit the Cu-Zn SOD.

The potential interferences with the xanthine oxidase-cytochrome *c* assay for SOD, already discussed, are rather obvious in both cause and in circumvention. There is a much more subtle problem which deserves notice although it is, happily, not frequently encountered. This arises when the assay mixture contains a substance which reacts rapidly with $O_2^{\cdot-}$ to give a radical able to react, in turn, with the indicating scavenger for $O_2^{\cdot-}$. Pamoic acid is such a substance and it was first studied as an apparent inhibitor of SOD, which it is not. Pamoate reacts with $O_2^{\cdot-}$ to give a pamoate radical and this can then reduce cytochrome *c*.[10] When pamoate is present the steady-state level of $O_2^{-\cdot}$ is markedly reduced, but the rate of cytochrome *c* reduction is unaffected. Pamoate thus appears not to interfere with the cytochrome *c* reduction assay. But, when SOD is added it is less effective than it would be in the absence of pamoate, because the pamoate then competes with the enzyme for the $O_2^{-\cdot}$ and because much of the cytochrome *c* reduction is then actually due to pamoate radical.

III. XANTHINE OXIDASE ASSAYS — VARIATIONS ON A THEME

Since the catalytic efficiency of the Cu-Zn SOD is independent of pH in the range 5 to 10,[11] whereas the rate of reaction of $O_2^{-\cdot}$ with cytochrome *c* decreases with increasing pH,[12-15] raising the pH gives a competitive advantage to SOD. This has been exploited and combined with a low concentration of cytochrome *c*, to provide an ultrasensitive assay for SOD.[16] A linear plotting form for the inhibition of cytochrome *c* reduction by SOD has been used to simplify the calculation of SOD concentration.[17,18]

$O_2^{-\cdot}$ can act both as a one-electron reductant and as a one-electron oxidant. Cytochrome *c* has the advantage that its one-electron reduction product is stable and its reaction with $O_2^{-\cdot}$ is thus stoichiometric and simple. Nitroblue tetrazolium (NBT) is often used as an indicating scavenger for $O_2^{-\cdot}$. Its reduction to the monoformazan requires two electrons and to the diformazan, four electrons. The reduction of NBT by $O_2^{-\cdot}$ must therefore involve unstable intermediates, the chemistry of which have been discussed by Bielski et al.[19] The existence of unstable intermediate reduced forms of NBT, such as the univalently reduced tetrazoinyl radical, raises the possibility that NBT can generate $O_2^{-\cdot}$, as well as scavenge $O_2^{-\cdot}$. Thus, given an enzyme capable of the direct univalent reduction of NBT, we could anticipate $O_2^{-\cdot}$ production by autoxidation of the tetrazoinyl radical. This effect has been observed with the NADPH-cytochrome P450 reductase.[20]

In the xanthine oxidase reaction, NBT has another drawback as an indicating scavenger for $O_2^{-\cdot}$ and that is its direct reduction by native xanthine oxidase. The reduction of NBT by xanthine oxidase is not oxygen dependent but, in air-equilibrated solutions, approximately 80% of the NBT reduction is mediated by $O_2^{-\cdot}$ and therefore is SOD inhibitable. It is necessary, in recognition of this, to define a unit of SOD as that amount which gives half of the maximum achievable inhibition. The one overwhelming virtue of NBT is that its stable reduction product is insoluble and this provides the basis for activity stains for SOD, applicable to polyacrylamide gel electropherograms,[21] which will be discussed subsequently. In doing solution assays with NBT the insolubility of the formazan is a problem which is partially overcome by the use of Triton® X-100.[22]

Indicating scavengers which are oxidized by $O_2^{-\cdot}$ have also been used as the basis for assaying SOD. Thus $O_2^{-\cdot}$ causes the first step leading to the oxidation of epinephrine to adrenochrome. For this reason the xanthine oxidase reaction causes a co-oxidation of epinephrine which is inhibitable by SOD.[4] The oxidation of NH_2OH to NO_2^- was similarly caused by $O_2^{-\cdot}$ and used as the basis of an assay for SOD,[23] as was the co-oxidation of sulfite.[24]

IV. ASSAYS BASED UPON PHOTOCHEMISTRY

Dyes, which are electronically excited by absorption of a photon, are more active oxidants than the same dyes in the ground state. If the dye can be reduced to an autoxidizable semiquinone, it will generate $O_2^{-\cdot}$ when mixed with a potential reductant and illuminated. The photoreduction of flavins by EDTA was noted by Frisell et al.[25] and was used by Massey et al.[26] and Ballou et al.[27] to provide a convenient source of $O_2^{-\cdot}$, which could be used to reduce cytochrome *c* and to assay SOD in terms of inhibition of this cytochrome *c* reduction. The riboflavin-sensitized photoproduction of $O_2^{\cdot -}$ was then used, with NBT as the indicating scavenger of $O_2^{\cdot -}$, to provide an assay for SOD which could be applied in free solution or to polyacrylamide gel electropherograms.[21] Because SOD intercepts the $O_2^{\cdot -}$ and thus prevents reduction of NBT to the purple formazan, regions bearing SOD are seen as achromatic zones against a purple background in this staining procedure.

Several variations of the photochemical NBT reduction assay for SOD have appeared. Thus Nishikimi et al.[28] used phenazine methosulfate in place of riboflavin and reduced nicotinamide adenine dinucleotide (NADH) in place of EDTA. Rapp et al.[29] used a light pipe to illuminate reaction mixtures at right angles to the analytical beam, thus allowing continuous monitoring of NBT reduction and providing a convenient photochemical solution assay for SOD. Riboflavin was seen to photosensitize the oxidation of dianisidine, and SOD augmented this process. This provided the basis for a photochemical assay in which SOD signaled its presence by increasing the rate of the reaction being observed, rather than by decreasing it, as is the case in the other assays. This "positive" or "augmentation" assay was developed for use in free solution[30] or on gel electropherograms.[31] The mechanism offered in explanation of this augmentation of dianisidine photooxidation involved the photoproduction of both $O_2^{\cdot -}$ and of a univalently oxidized dianisidine. The dianisidine radical could then either dismute to the colored product, or be reduced back to dianisidine by the $O_2^{\cdot -}$. In the absence of SOD the back reduction predominated and thus masked the true extent of the photochemical process.[30]

V. SUPEROXIDE-PROPAGATED CHAIN REACTIONS

Numerous autoxidations are free-radical chain reactions in which $O_2^{\cdot -}$ figures as a chain propagating species. In such situations SOD will, by scavenging the $O_2^{\cdot -}$, shorten the reaction chains and so decrease the overall rate of that autoxidation. The autoxidation of epinephrine in the alkaline pH range is inhibited by SOD and this has been used as the basis for an assay for SOD.[32] Pyrogallol has the advantage of autoxidizing readily at pH 8, whereas with epinephrine pH values above 9.0 were required. SOD has been assayed on the basis of its ability to inhibit the autoxidation of pyrogallol[33] or of 6-hydroxydopamine.[34] The autoxidation of NH_2OH at elevated pH has also been developed as the basis of an assay for SOD.[35] The first of the autoxidation assays, that based upon epinephrine, has been improved by following the process in the near UV, rather than in the visible.[36]

Autoxidation assays have the virtue of simplicity. A mildly acid, and therefore stable, stock solution of the substrate is delivered into the alkaline buffer and the rate of autoxidation followed spectrophotometrically, fluorometrically or with a Clark oxygen electrode. As with other assays, one unit of SOD is defined as that amount causing 50% inhibition of the rate. Free-radical chain reactions are however notoriously subject to initiation by transition metals and to inhibition by antioxidants. They are therefore intrinsically more prone to interferences than are other types of assays. They are nevertheless very useful when many partially purified samples must be assayed, as in assaying column eluants.

VI. DIRECT ASSAYS

We stated at the outset that $O_2^{\cdot -}$ was too unstable to lend itself to the traditional direct observation of the relevant enzymic reaction. However, if the time of observation is dramatically shortened, or if one works at sufficiently elevated pH, direct assays are possible. Pulse radiolysis, in which a beam of electrons is directed into the reaction mixture, allows the introduction, on a microsecond time scale, of $\sim 10^{-4} M\ O_2^{\cdot -}$. The disappearance of this $O_2^{\cdot -}$ can then be followed, on a millisecond time scale, by direct UV spectrophotometry. The catalytic activity of SOD can be directly measured in this way.[11] The equipment required is too expensive and too uncommonly available for this to be useful as a routine means of assaying SOD. It has however been important as a means of validating the indirect assays and of studying the mechanisms of the SODs.[11,37-41] Marklund[42] used elevated pH to slow the spontaneous dismutation sufficiently to allow its direct observation on a time scale of minutes. He use commercially available KO_2 as the source of $O_2^{\cdot -}$ and found that he could readily assay as little as 1 $\times 10^{-10} M$ SOD and that this assay could be applied to tissue extracts.

VII. SPIN TRAPPING ASSAYS

The direct assays just described require instantaneous concentrations of $O_2^{\cdot -}$ which are high enough to observe directly, either by UV spectrophotometry[43] or by ESR.[44] Because the spontaneous dismutation of $O_2^{\cdot -}$ is a second-order process, such high concentrations are hard to achieve. The indirect methods, in which $O_2^{\cdot -}$ converts an indicating scavenger into a stable product, are integrative and can operate at very low steady-state levels of $O_2^{\cdot -}$. Spin trapping is another integrative method in which the unstable free radical, $O_2^{\cdot -}$, reacts with some scavenger yielding a stable free radical, which accumulates and which can then be measured by ESR. These methods have the advantage of being qualitative as well as quantitative. Thus the nature of the radical trapped as well as its amount can be deduced from the ESR signal.[45-48]

VIII. DISTINGUISHING ONE TYPE OF SOD FROM ANOTHER

There are three classes of SODs, i.e., those based upon Cu and Zn, upon Mn, or upon Fe. All of these are comparably active at neutral pH, yet it is often desirable to be able to assay them separately in mixtures. Several means of doing this have been developed. Thus Cu and Zn SODs are resistant to the Tsuchihashi treatment with $CHCl_3$ and ethanol, but are inhibited by millimolar levels of cyanide, whereas the MnSOD is destroyed by the Tsuchihashi treatment but is unaffected by cyanide. This allows these enzymes to be distinguished.[49] The Cu-Zn SOD[50,51] and the Fe SOD[52] can be inactivated by treatment with H_2O_2 under conditions which do not affect the Mn SOD. The Cu-Zn SOD is as active at pH 10 as it is at pH 7.8, whereas the Mn SOD is less active at the elevated pH.[53] The different types of SOD show different sensitivity to inhibition by azide.[54] All of these methods have been successfully applied to distinguishing among these members of the SOD family, but since they depend upon incidental properties of the enzymes they could ultimately prove misleading.

SODs of all classes have been successfully resolved in a reversible fashion. That is, means have been found for reversibly removing the metal prosthetic group. In all cases the apoprotein was inactive and could be reactivated only by restoring the metal native to that particular SOD. Thus only Cu would restore activity to the apoCu-Zn SOD, only Fe to the apoFe SOD and only Mn to the apoMn SOD. Reversible resolution has been used to distinguish one member of the SOD family from another,[55] and it is intrinsically more reliable, since it depends upon the structure of the active site region.

Radioimmunoassays, although laborious, offer both great sensitivity and great specificity for a particular SOD and several such assays have been developed. Kelly et al.[56] used a rabbit anti-bovine Cu-Zn SOD as the basis of a double antibody solid phase radioimmunoassay. Comparable assays have been developed for the Cu-Zn SODs from humans[57] and several other sources.[58] A radioimmunoassay for the human Mn SOD has also been reported.[59]

IX. MISCELLANEOUS ASSAYS

The cathodic reduction of oxygen generates $O_2^{\cdot-}$ which then diffuses from the electrode surface.[60] If the dismutation reaction occurs close to that surface, it generates O_2 which can diffuse back to the cathode and accept another electron. In essence, the dismutation of $O_2^{\cdot-}$ close to the electrode has the same effect as increasing the pO_2 in the bulk solution would have, i.e., it increases the rate at which O_2 diffuses to the surface of the electrode. This has been used to develop a polarographic assay for SOD in which the height of the oxygen reduction wave at a dropping mercury cathode is increased by SOD.[61] In this method the pH was kept alkaline (pH 9.9) with a borate buffer and a hydrophobic layer was formed on the cathode by including triphenyl phosphine oxide. In a later development this polarographic method was adapted to the measurement of both SOD and catalase.[62] These methods appear to be sensitive enough for most purposes but, because of a large baseline, they will inevitably lack precision.

SOD has been reported to enhance the longitudinal nuclear magnetic relaxation rate of ^{19}F and this has been made the basis of an NMR assay for SOD.[63] Both Cu-Zn SOD and Mn SOD were reported to have this effect on the ^{19}F NMR, while other metalloproteins did not. This is an extremely unusual assay, in that it does not reflect the catalytic activity or the specific antigenic features of the enzymes being assayed. Presumably it reflects some peculiarity of the ligand field of the active site metals and may be useful to those already doing NMR.

REFERENCES

1. **Crapo, J. D., McCord, J. M., and Fridovich I.,** Preparation and assay of superoxide dismutases, *Meth. Enzymol.*, 53, 382, 1978.
2. **Bielski, B. H. J. and Allen, A. O.,** Mechanism of the disproportionation of superoxide radicals, *J. Phys. Chem.*, 81, 1048, 1977.
3. **McCord, J. M. and Fridovich, I.,** The reduction of cytochrome *c* by milk xanthine oxidase, *J. Biol. Chem.*, 243, 5753, 1968.
4. **McCord, J. M. and Fridovich, I.,** Superoxide dismutase — an enzymic function for erythrocuprein, *Fed. Proc.*, 28, 346, 1968.
5. **McCord, J. M. and Fridovich, I.,** Superoxide dismutase — an enzymic function for erythrocuprein (hemocuprein), *J. Biol. Chem.*, 244, 6049, 1969.
6. **Fridovich, I.,** Quantitative aspects of the production of superoxide anion radical by milk xanthine oxidase, *J. Biol. Chem.*, 245, 4053, 1970.
7. **Fridovich, I. and Handler, P.,** Xanthine oxidase. V. Differential inhibition of the reduction of various electron acceptors, *J. Biol. Chem.*, 237, 916, 1962.
8. **Waud, W. R., Brady, F. O., Wiley, R. D., and Rajagopalan, K. V.,** A new purification procedure for bovine milk xanthine oxidase: effect of proteolysis on the subunit structure, *Arch. Biochem. Biophys.*, 169, 695, 1975.
9. **Halliwell, B.,** Superoxide dismutase: a contaminant of bovine catalase, *Biochem. J.*, 135, 379, 1973.
10. **Hassan, H. M., Dougherty, H., and Fridovich, I.,** Inhibitors of superoxide dismutases: a cautionary tale, *Arch. Biochem. Biophys.*, 199, 349, 1980.
11. **Klug, D., Rabani, J., and Fridovich, I.,** A direct demonstration of the catalytic action of superoxide dismutase through the use of pulse radiolysis, *J. Biol. Chem.*, 247, 4839, 1972.
12. **Butler, J., Jayson, G. G., and Swallow, A. J.,** The reaction between the superoxide anion radical and cytochrome *c*, *Biochim. Biophys. Acta*, 408, 215, 1975.
13. **Simic, M. G., Taub, T. A., Tacci, J., and Hurwitz, P. A.,** Free radical reduction of ferricytochrome *c*, *Biochem. Biophys. Res. Commun.*, 62, 161, 1975.
14. **Koppenol, W. H., Van Buuren, K. J. H., Butler, J., and Braams, R.,** The kinetics of the reduction of cytochrome *c* by the superoxide anion radical, *Biochim. Biophys. Acta*, 449, 157, 1976.

15. **Seki, H., Ilan, Y. A., Ilan, Y., and Stein, G.,** Reactions of the ferri-ferrocytochrome-*c* system with superoxide/oxygen and $CO_2^{\overline{\cdot}}/CO_2$ studied by fast pulse radiolysis, *Biochim. Biophys. Acta,* 440, 573, 1976.
16. **Salin, M. L. and McCord, J. M.,** Superoxide dismutase in polymorphonuclear leukocytes, *J. Clin. Invest.,* 54, 1005, 1974.
17. **Sawada, Y. and Yamazaki, I.,** One-electron transfer reactions in biochemical systems. VIII. Kinetic study of superoxide dismutase, *Biochim. Biophys. Acta,* 327, 257, 1973.
18. **Nishikimi, M.,** Oxidation of ascorbic acid with superoxide anion generated by the xanthine-xanthine oxidase system, *Biochem. Biophys. Res. Commun.,* 63, 463, 1975.
19. **Bielski, B. H. J., Shive, G. G., and Bajuk, S.,** Reduction of nitro blue tetrazolium by CO_2^- and $O_2^{\overline{\cdot}}$ radicals, *J. Phys. Chem.,* 84, 830, 1980.
20. **Auclair, C., Torres, M., and Hakim, J.,** Superoxide anion involvement in NBT reduction catalyzed by NADPH-cytochrome P-450 reductase: a pitfall, *FEBS Lett.,* 89, 26, 1978.
21. **Beauchamp, C. and Fridovich, I.,** Superoxide dismutase: improved assays and an assay applicable to polyacrylamide gels, *Anal. Biochem.,* 44, 276, 1971.
22. **Nishikimi, M.,** The generation of superoxide anion in the reaction of tetrahydropteridines with molecular oxygen, *Arch. Biochem. Biophys.,* 166, 273, 1975.
23. **Elstner, E. F. and Heupel, A.,** Inhibition of nitrite formation from hydroxyl ammonium chloride: a simple assay for superoxide dismutase, *Anal. Biochem.,* 70, 616, 1976.
24. **Tyler, B. B.,** Polarographic assay and intracellular distribution of superoxide dismutase in rat liver, *Biochem. J.,* 147, 493, 1975.
25. **Frisell, W. R., Chung, C. W., and Mackenzie, C. G.,** Catalysis of oxidation of nitrogen compounds by flavin coenzymes in the presence of light, *J. Biol. Chem.,* 234, 1297, 1959.
26. **Massey, V., Strickland, S., Mayhew, S. G., Howell, L. G., Engel, P. C., Matthews, R. G., Schuman, M., and Sullivan, P. A.** The production of superoxide anion radicals in the reaction of reduced flavins and flavoproteins with molecular oxygen, *Biochem. Biophys. Res. Commun.,* 36, 891, 1969.
27. **Ballou, D., Palmer, G., and Massey, V.,** Direct demonstration of superoxide anion production during the oxidation of reduced flavin and of its catalytic decomposition by erythrocuprein, *Biochem. Biophys. Res. Commun.,* 36, 898, 1969.
28. **Nishikimi, M., Rao, N. A., and Yagi, K.,** The occurrence of superoxide anion in the reaction of reduced phenazine methosulfate and molecular oxygen, *Biochem. Biophys. Res. Commun.,* 46, 849, 1972.
29. **Rapp, U., Adams, W. C., and Miller, R. W.,** Purification of superoxide dismutase from fungi and characterization of the reaction of the enzyme with catechols by electron spin resonance spectroscopy, *Can. J. Biochem.,* 51, 158, 1973.
30. **Misra, H. P. and Fridovich, I.,** Superoxide dismutase: a photochemical augmentation assay, *Arch. Biochem. Biophys.,* 181, 308, 1977.
31. **Misra, H. P. and Fridovich, I.,** Superoxide dismutase and peroxidase: a positive activity stain applicable to polyacrylamide gel electropherograms, *Arch. Biochem. Biophys.,* 183, 511, 1977.
32. **Misra, H. P. and Fridovich, I.,** The role of superoxide anion in the autoxidation of epinephrine and a simple assay for superoxide dismutase, *J. Biol. Chem.,* 247, 3170, 1972.
33. **Marklund, S. and Marklund, G.,** Involvement of the superoxide anion radical in the autoxidation of pyrogallol and a convenient assay for superoxide dismutase, *Eur. J. Biochem.,* 47, 469, 1974.
34. **Heikkila, R. E. and Cabbat, F.,** A sensitive assay for superoxide dismutase based on the autoxidation of 6-hydroxydopamine, *Anal. Biochem.,* 75, 356, 1976.
35. **Kono, Y.,** Generation of superoxide radical during autoxidation of hydroxylamine and an assay for superoxide dismutase, *Arch. Biochem. Biophys.,* 186, 189, 1978.
36. **Sun, M. and Zigman, S.,** An improved spectrophotometric assay for superoxide dismutase based on epinephrine autoxidation, *Anal. Biochem.,* 90, 81, 1978.
37. **Klug, D., Fridovich, I., and Rabani, J.,** Pulse radiolytic investigation of superoxide dismutase-catalyzed disproportionation. I. Mechanism for bovine superoxide dismutase, *J. Am. Chem. Soc.,* 95, 2786, 1973.
38. **Rotilio, G., Bray, R. C., and Fielden, E. M.,** A pulse radiolysis study of superoxide dismutase, *Biochim. Biophys. Acta,* 268, 605, 1972.
39. **Fielden, E. M., Roberts, P. B., Bray, R. C., and Rotilio, G.,** Mechanism and inactivation of superoxide dismutase activity, *Biochem. Soc. Trans.,* 1, 52, 1973.
40. **Pick, M., Rabani, J., Yost, F., and Fridovich, I.,** The catalytic mechanism of the manganese-containing superoxide dismutase of *E. coli* studied by pulse radiolysis, *J. Am. Chem. Soc.,* 96, 7329, 1974.
41. **McAdam, M. E., Fox, R. A., Lavelle, F., and Fielden, E. M.,** A pulse-radiolysis study of the manganese-containing superoxide dismutase from *Bacillus stearothermophilus*; and Further studies on the properties of the enzyme, *Biochem. J.,* 165, 71 and 81, 1977.

42. **Marklund, S.**, Spectrophotometric study of spontaneous disproportionation of superoxide anion radical and sensitive direct assay for superoxide dismutase, *J. Biol. Chem.*, 251, 7504, 1976.
43. **Bielski, B. H. J.**, Reevaluation of the spectral and kinetic properties of HO_2 and O_2^- free radicals, *Photochem. Photobiol.*, 28, 645, 1978.
44. **Knowles, P. F., Gibson, J. F., Pick, F. M., and Bray, R. C.**, Electron-spin-resonance evidence for enzymic reduction of oxygen to a free radical, the superoxide ion, *Biochem. J.*, 111, 53, 1969.
45. **Finkelstein, E., Rosen, G. M., Rauckman, E. J., and Paxton, J.**, Spin trapping of superoxide, *Mol. Pharmacol.*, 16, 676, 1979.
46. **Finkelstein, E., Rosen, G. M., and Rauckman, E. J.**, Spin trapping: kinetics of the reaction of superoxide and hydroxyl radicals with nitrones, *J. Am. Chem. Soc.*, 102, 4994, 1980.
47. **Finkelstein, E., Rosen, G. M., and Rauckman, E. J.**, Spin trapping of superoxide and hydroxyl radical: practical aspects, *Arch. Biochem. Biophys.*, 200, 1, 1980.
48. **Finkelstein, E.**, Studies on the Biological Detection, Production and Fate of Superoxide, Ph.D. thesis, Duke University, Durham, N.C., 1980.
49. **Weisiger, R. A. and Fridovich, I.**, Superoxide dismutase: organelle specificity, *J. Biol. Chem.*, 248, 3582, 1973.
50. **Bray, R. C., Cockle, S. A., Fielden, E. M., Roberts, P. B., Rotilio, B., and Calabrese, L.**, Reduction and inactivation of superoxide dismutase by hydrogen peroxide, *Biochem. J.*, 139, 43, 1974.
51. **Hodgson, E. K. and Fridovich, I.**, The interaction of bovine erythrocyte superoxide dismutase with hydrogen peroxide: inactivation of the enzyme, *Biochemistry*, 14, 5294, 1975.
52. **Asada, K., Yoshikawa, K., Takahashi, M. -A., Maeda, Y., and Enmanji, K.**, Superoxide dismutases from a blue-green alga, *Plectonema boryanum*, *J. Biol. Chem.*, 250, 2801, 1975.
53. **Salin, M., Day, E. D., and Crapo, J. D.**, Isolation and characterization of a manganese-containing superoxide dismutase from rat liver, *Arch. Biochem. Biophys.*, 187, 223, 1978.
54. **Misra, H. P. and Fridovich, I.**, Inhibition of superoxide dismutase by azide, *Arch. Biochem. Biophys.*, 189, 317, 1978.
55. **Kirby, T., Blum, J., Kahane, I., and Fridovich, I.**, Distinguishing between Mn containg and Fe containing superoxide dismutases in crude extracts of cells, *Arch. Biochem. Biophys.*, 201, 551, 1980.
56. **Kelly, K., Barefoot, C., Sehon, A., and Petkau, A.**, Bovine superoxide dismutase: a radioimmunoassay, *Arch. Biochem. Biophys.*, 190, 531, 1978.
57. **Del Villano, B. C. and Tischfield, J. A.**, Radioimmune assay for human superoxide dismutase in erythrocytes, *Fed. Proc.*, 37, 1749, 1978.
58. **Baret, A., Michel, P., Imbert, M. R., Morcellet, J. L., and Michelson, A. M.**, A radioimmunoassay for copper containing superoxide dismutase, *Biochem. Biophys. Res. Commun.*, 88, 337, 1979.
59. **Baret, A., Schiavi, P., Michel, P., Michelson, A. M., and Puget, K.**, A radioimmunoassay for manganese containing superoxide dismutase, *FEBS Lett.*, 112, 25, 1980.
60. **Forman, H. J. and Fridovich, I.**, The electrolytic univalent reduction of oxygen in aqueous solution demonstrated through the use of superoxide dismutase, *Science*, 175, 339, 1972.
61. **Rigo, A., Viglino, P., and Rotilio, G.**, Polarographic determination of superoxide dismutase, *Anal. Biochem.*, 68, 1, 1975.
62. **Rigo, A. and Rotilio, G.**, Simultaneous determination of superoxide dismutase and catalase in biological materials by polarography, *Anal. Biochem.*, 81, 157, 1977.
63. **Rigo, A., Viglino, P., Argese, E., Terenzi, M., and Rotilio, G.**, Nuclear magnetic relaxation of ^{19}F as a novel assay method of superoxide dismutase, *J. Biol. Chem.*, 254, 1759, 1979.

Chapter 4

OXYGEN TOXICITY IN PROKARYOTES: THE IMPORTANCE OF SUPEROXIDE DISMUTASE

Irwin Fridovich

TABLE OF CONTENTS

I. INTRODUCTION

The incredible diversity of life forms is underpinned by a biochemistry of remarkable constancy. Experience has shown that a great deal may be learned about the molecular biology and metabolism of higher life forms from studies of the lower forms. There will of course be differences and there may even be surprises, so that the higher organisms must ultimately be examined directly. Nevertheless, an expeditious approach to a biochemical problem begins with the simpler organisms, because they are so much easier to manipulate and because the understanding gained from them guides the experimental approaches to the more complex organisms. The dictum then is: study *E. coli* first, elephants later. In this chapter we will discuss studies of oxygen toxicity in prokaryotes, with the clear understanding that what we learn from *E. coli* will probably be applicable to elephants and to man.

II. SURVEYS: THE SIMPLEST APPROACH

The discovery of superoxide dismutase (SOD)[1-3] led naturally and rapidly to the proposal that $O_2^{-\cdot}$ was an important agent of the toxicity of oxygen and that SOD was an essential defense. The simplest and first exploration of this proposal was a survey of a variety of microorganisms for their contents of SOD and of catalase. The expectation was that organisms which presumably never encountered dioxygen could do without SOD, whereas those which utilized O_2 would need this defense. The results were generally in accord with this expectation.[4] Thus, facultative aerobes contained both SOD and catalase, while aerotolerant anaerobes had less SOD and occasionally no catalase. The sensitive obligate anaerobes tested had neither SOD nor catalase.

These results were encouraging for the superoxide theory of oxygen toxicity, but the situation with anaerobes is more complex than was revealed in this first study. Thus some anaerobes have since been found to contain SOD.[5-15] On a well-oxygenated planet this should not be surprising. Even organisms which habitually live in anaerobic environments need to tolerate temporary oxygen exposure, en route from one anaerobic niche to the next. Besides, some environments which are traditionally considered anaerobic, such as the intestinal tracts of larger animals, are not really oxygen free.[16] Some of the anaerobes, which have been reported to contain SOD, contain very small amounts. For example, *Clostridium acetobutylicum* and *Cl. pasteurianum* have only 1 to 6% as much SOD as do aerobically grown *E. coli.*[6] There is actually a wide range of oxygen-tolerance among anaerobes[7] and in a collection of anaerobes the oxygen tolerance and SOD content varied in parallel.[10,12] In the anaerobe *Selenomonas ruminantium* levels of oxygen which inhibited growth also caused increases in SOD.[14] A similar observation was made with *Bacterioides fragilis,*[17] in which case elevated SOD imparted resistance towards 5 atm O_2.

III. *LACTOBACILLUS PLANTARUM*

SOD thus seems to serve the same purpose in anaerobes as it does in aerobes, which is to provide a defense against oxygen toxicity. The existence of this enzyme in anaerobes is not inconsistent with the superoxide theory of oxygen toxicity. *L. plantarum* did, however, appear to be an embarrassment for this theory. Thus *L. plantarum,* although aerotolerant, was reported to be devoid of SOD.[4] This was at first satisfactorily rationalized by the observation that *L. plantarum,* grown on a glucose medium, did not respire.[4,18] An organism which did not reduce O_2 could not make $O_2^{-\cdot}$ and so could afford to dispense with SOD, even when bathed in an oxygenated medium.

This explanation does not hold up under close scrutiny. *L. plantarum* does consume

O_2 under certain conditions.[19] Even cells grown on glucose were observed to develop some respiration during stationary phase in glucose-exhausted media.[20] A close examination of the life and times of *L. plantarum* has revealed that this organism has replaced SOD with another scavenger of $O_2^{\overline{\cdot}}$ and this is Mn(II). While most organisms protect themselves against $O_2^{\overline{\cdot}}$ with micromolar levels of SOD, *L. plantarum* achieves the same end with millimolar levels of Mn(II) and it does exhibit oxygen-intolerance when deprived of Mn(II).[21] Why *L. plantarum* has chosen this unusual strategy is not presently clear, but its need for a defense against $O_2^{\overline{\cdot}}$ is a strong endorsement of the importance of this radical in oxygen toxicity.

IV. INDUCTIONS OF SOD

Maximum efficiency is achieved by modulating the amount of any particular enzyme to match the need for its services. When a substrate is not being handled, there is no need to waste energy and substance in synthesizing the enzymes which use that substrate and its metabolic derivatives. Thus, when yeast cells are growing on glucose they have no need for the enzymes of galactose uptake and metabolism and they do not make them. When presented with galactose or galactosides instead of glucose, they begin to make and accumulate the required enzymes. Induction of the biosynthesis of an enzyme by a particular substrate is probably the surest indication that its in vivo function is to act upon that substrate or upon metabolic derivatives thereof. For facultative organisms O_2, and therefore $O_2^{\overline{\cdot}}$, is a "sometime thing" and we should expect to see at least some modulation of the level of SOD to the needs of the moment.

V. MODULATION OF DEFENSIVE ENZYMES

Enzymes which provide a defense cannot be modulated quite as aggressively as enzymes which process a carbon source. Thus a cell faced with a galactoside in the absence of the appropriate permease and the β-galactosidase pays only a modest price for its lack of preparedness and that is an inability to grow for the span of minutes needed to synthesize these enzymes. In contrast, a cell faced with O_2 and lacking SOD could succumb to oxygen toxicity before it had a chance to make the needed enzyme. The level of the galactose enzymes may thus be allowed to fall to undetectably low levels in the absence of galactose, while the SOD can only be allowed to vary within narrower limits. Nevertheless, very clear and sometimes dramatic inductions of SOD have been seen, and in all cases examined, increased intracellular levels of SOD correlated with increased resistance towards oxygen.

Induction of SOD caused by increased exposure to O_2 has been seen in *Streptococcus faecalis* and *E. coli*,[22-24] in rat lung,[25-31] yeast,[32] *Photobacter leiognathi*,[33] leukocytes,[34-37] rat mammary carcinoma,[38] *Propionibacterium shermanii*,[39] potato slices,[40] *B. fragilis*,[17] *Vibrio eltor*,[41] and *Oscillatoria limnetica*.[42] With *S. faecalis* and *E. coli*, increased levels of SOD achieved by oxygen exposure correlated with increased resistance towards oxygen toxicity[22-24] and towards the oxygen enhancement of the toxicity of streptonigrin.[23] These results support the superoxide theory of oxygen toxicity, but suffer from the circularity involved in inducing the SOD with O_2 and then noting a gain in resistence towards this same stress. Our argument would be better served if we could manipulate the level of SOD without changing pO_2 and then see a correlation between enzyme level and oxygen tolerance. This has been achieved by several different strategies.

VI. *E. COLI* K12: RESPONSIVE TO ULTRA LOW pO_2

E. coli B and *E. coli* K12 were, for a time, thought to differ qualitatively in their response to oxygen.[32] Thus, transfer from deep, static liquid culture to forced aeration occasioned an induction of SOD in the case of *E. coli* B, but not of *E. coli* K12. It was assumed that deep, static liquid culture was effectively anaerobic culture, even though the flasks were in communication with the atmosphere through cotton plugs, and thus that *E. coli* B could respond to O_2 by induction of SOD, whereas *E. coli* K12 could not.

In fact, *E. coli* K12 can induce SOD in response to O_2, but is so sensitive that it is essentially fully induced by the low levels of O_2 present in the quiescent liquid cultures,[43] whereas *E. coli* B is less responsive to low levels of O_2.

When grown under truly anaerobic conditions *E. coli* K12 contains only an iron-containing SOD (Fe SOD), but when exposed to O_2, even the miniscule amounts in a deep static liquid culture, it induces a manganese-containing SOD (Mn SOD).[43] *E. coli* thus solves the dilemma inherent in modulating a protective enzyme. It contains two isozymes, one of which is aggressively modulated, while the other is not modulated at all and so provides a back-up defense. The constitutive Fe SOD thus prevents rapid death of the cells upon sudden exposure to O_2 and so buys the time needed for synthesis of the Mn SOD. Induction of the Mn SOD correlates with increased resistance to oxygen toxicity in *E. coli* K12, as it does in *E. coli* B.

VII. CHEMOSTAT: INDUCTION AS A FUNCTION OF GROWTH RATE

When grown in a glucose-limited, vigorously aerated, chemostat, *E. coli* responded to increased influx of fresh medium with enhanced growth and increased respiration.[44] Under this condition of constant aeration, increased growth rate also caused increased intracellular levels of SOD and increased levels of SOD, in turn, correlated with resistance of the cells towards the lethality of hyperbaric oxygen. Exposures to hyperbaric oxygen were done in fresh medium containing chloramphenicol to prevent induction of SOD during the oxygen exposure. Exposure to hyperbaric N_2 under identical conditions caused no loss of viability. In this case induction of SOD had been achieved under conditions of constant aeration, so there was no circularity in the final correlation of elevated SOD with enhanced resistance towards oxygen toxicity.

These experiments also uncovered an intriguing control on the rate of growth by the level of SOD, which is not yet understood. Thus increased access to fresh medium, during chemostat culture, caused increased growth rate, increased respiration, and increased levels of SOD. What would happen if fresh medium was added suddenly to cells whose growth had been nutritionally limited in the chemostat? Would they suddenly increase their rates of growth and of respiration? When this nutritional step-up experiment was performed,[44] the culture continued to grow at the slow rate, characteristic of the nutritional limitation, for 1 hr and only then switched over to a more rapid rate of growth. During that hour of lagging growth rate the level of SOD within the cells was increased and only when it reached the concentration appropriate to the higher growth rate did the growth rate increase.

VIII. THE GLUCOSE EFFECT

When *E. coli* was grown on a trypticase soy yeast extract (TSY) medium, its content of SOD was seen to vary in a reproducible way during the growth cycle. Thus, during the first few hours of growth, the SOD content fell to a minimum and then increased again.[45] While the SOD level was falling, the glucose content and the pH of the medium both fell, indicating glucose fermentation. When the glucose was exhausted, the cells began to utilize the accumulated organic acids and the amino acids in the medium and both pH and SOD content rose again. This effect of glucose was not due to classical catabolite repression, since 3′,5′-cyclic adenylic acid (cyclic AMP) had no effect on the variation of SOD content during the growth cycle and since α-methyl glucoside, which does cause catabolite repression, did not act like glucose in suppressing the synthesis of SOD.[45]

It appears that *E. coli* will preferentially utilize glucose, when it is available, and will then rely on fermentation to satisfy their energy needs. When the glucose content of the TSY medium was exhausted, the cells switched over to an aerobic metabolism and presumably then generated more $O_2^{\overline{\cdot}}$, which induced the synthesis of more SOD. This supposition was tested by artificially increasing the intracellular production of $O_2^{\overline{\cdot}}$ with methyl viologen (paraquat). When this was done there was a profound induction of SOD, which was seen even in fresh TSY medium containing glucose.[45]

IX. INDUCTIONS BY PARAQUAT AND RELATED COMPOUNDS

There are compounds which can enter *E. coli* and other cells and there divert electron flow from the water-producing cytochrome pathway to an $O_2^{\overline{\cdot}}$-producing pathway. These compounds are rapidly reduced by diaphorases in the cells, yielding reduced forms which are rapidly reoxidized by O_2, yielding $O_2^{\overline{\cdot}}$. Paraquat, pyocyanine, juglone, and plumbagin are among the compounds which function in this way. They all act catalytically to increase $O_2^{\overline{\cdot}}$ production, and since they provide a cytochrome-independent path to oxygen, to increase cyanide-resistant respiration,[46-49] these compounds exhibit an oxygen-dependent toxicity. In a real sense they exacerbate the toxicity of oxygen. If their effect was really due to increased intracellular production of $O_2^{\overline{\cdot}}$, then an electron source should be as indispensible for their action as is O_2. Thus, $O_2^{\overline{\cdot}}$ can be made only if both O_2 and a source of electrons is present. In fact, the lethality of paraquat does depend entirely upon both O_2 and upon a rapidly metabolizable substrate.

E. coli were found to be much more tolerant of paraquat when suspended in a rich medium, such as TSY, than in a glucose-salts medium.[47] This was due to the rapid induction of SOD in the rich medium, which was not possible in the minimal medium. When rapid induction of SOD was prevented with puromycin, the protective effect of the rich medium was eliminated. Since there are several ways to manipulate intracellular SOD in bacteria, it was possible to demonstrate directly[47] that elevated SOD did impart resistance towards the lethality of paraquat. Induction by paraquat or by related compounds, like induction by O_2, is due to increased synthesis of the Mn SOD, the Fe SOD being unaffected. The similarity of the actions of compounds such as paraquat, juglone, pyocyanine, and streptonigrin is emphasized by the finding that induction of SOD by one imparts resistance to the others.

Redox active compounds, which increase intracellular $O_2^{\overline{\cdot}}$ production, constitute a new group of antibiotics, whose action involves exacerbation of oxygen toxicity. Their effects demonstrate that $O_2^{\overline{\cdot}}$ is an important agent of oxygen toxicity and that SOD, which catalytically scavenges this radical, provides an essential defense. The components of the electron transport chain are not readily autoxidized. Nevertheless, they do autoxidize to some degree, which accounts for the small production of $O_2^{\overline{\cdot}}$ and of H_2O_2 by submitochondrial particles.[50-56] Blocking cytochrome oxidase with cyanide should increase the degree of steady-state reduction of these electron carriers and should therefore increase the rate at which they reduce O_2 to $O_2^{\overline{\cdot}}$. This leads to the expectation that cyanide should itself induce SOD. It has been shown to do so.[48]

X. EFFECTS OF EXOGENOUS $O_2^{\overline{\cdot}}$

When paraquat is reduced in *E. coli* it can reoxidize *in situ*, or can diffuse from the cell and then oxidize in the medium. In the former case it produces $O_2^{\cdot-}$ within the cell and in the latter case outside of the cell. In the complete absence of oxygen the reduced, blue, monocation radical of paraquat can be seen to accumulate in the suspending medium. Since the rate of reaction of the paraquat radical with O_2 is very rapid,[57] most of the paraquat radical will reoxidize within the cell under aerobic conditions. However, rapidly respiring *E. coli* can significantly deplete intracellular O_2, even in an aerobic medium, and some paraquat radical will then diffuse from the cell. Under such conditions part of the lethal effect of the paraquat is due to extracellular $O_2^{\cdot-}$ and H_2O_2 and some protection is afforded by SOD and by catalase, added to the suspending medium.[58]

There have been other reports of the lethality of exogenous $O_2^{\cdot-}$. *E. coli* B, grown in an iron-deficient medium, were killed by a photochemical source of $O_2^{\cdot-}$ and SOD added to the suspending medium provides essentially complete protection.[24] Similar results were obtained with an enzymic source of $O_2^{\cdot-}$.[24] *Lactobacillus plantarum* was killed by a photochemical source of $O_2^{\cdot-}$, and protection by both SOD and by catalase was noted.[18] There have also been conflicting reports. Both *E. coli* and *Staphylococcus epidermidis* were killed by the $O_2^{\cdot-}$ produced by the xanthine oxidase reaction. *S. epidermidis* was protected by both SOD and by catalase, while *E. coli* was significantly protected only by catalase.[59] In the case of *Neisseria gonorrhoeae*, catalase protected against the lethality of the xanthine oxidase reaction and SOD did not.[60] *S. albus* were killed by negative air ions (presumably $O_2^{\cdot-}$), generated in the gas phase and directed at a grounded suspension of the cells and SOD protected completely, whereas catalase was without effect.[61] SOD, added to the medium, greatly increased the oxygen resistance of *Campylobacter fetus*, while catalase was less effective.[62]

In some of these reported instances, $O_2^{\cdot-}$ appears to have been the important lethal agent, while in others both $O_2^{\cdot-}$ and H_2O_2 were involved and in still others H_2O_2 was itself the important factor. Some of these apparent conflicts may be explained on the basis of differences in the test organism or in the $O_2^{\cdot-}$-generating system. Rosen and Klebanoff[63] have performed a careful study of some of these factors. Thus they noted that exposure to the xanthine oxidase reaction killed *Staphylococcus aureus* and *Sarcina lutea* and that both $O_2^{\cdot-}$ and H_2O_2 were involved because both SOD and catalase protected. There were strain differences, as seen in the much greater susceptibility of a nonpigmented, compared to a yellow, *S. lutea*. There were also differences based on the $O_2^{\cdot-}$-generating system, in that lethality was greater with CH_3CHO than with xanthine, as the substrate for xanthine oxidase. This was apparently the case because the purine substrate, and to an even greater degree, the purine product, urate, protected the cells against the lethality of the system. Finally, if myeloperoxidase plus a halide were also present, the lethality of the reaction mixture was greatly augmented and then depended primarily upon H_2O_2. It does seem clear that in at least some circumstances, exogenous $O_2^{\cdot-}$ can exert marked deleterious effect on bacteria. This statement can be made for viruses as well.[64]

XI. EFFECTS OF RADIOGENIC O_2

If $O_2^{\cdot-}$ generated in the suspending medium, by enzymic or photochemical processes, can harm bacteria, what about $O_2^{\cdot-}$ generated by ionizing radiation? SOD added to the medium has been reported to protect myoblasts against such irradiation,[65] and mice have been protected against whole body exposures by injected SOD.[66] In suspensions of *E. coli* exposed to X-irradiation the lethality was enhanced 2.4-fold by the presence of oxygen and both SOD and catalase decreased this oxygen enhancement.[67] The diminution of the oxygen effect by SOD in bacterial suspensions was repeated independently,[68,69] but this effect was not seen with T_4 bacteriophage.[70] More work will be needed to sort out the conditions under which $O_2^{\cdot -}$ is an important factor in the oxygen enhancement of radiation lethality.

XII. SOD MUTANTS

A powerful way to clarify the physiological significance of SOD would be to create mutants in this enzyme and to compare the behavior of such mutants with that of the parental strain. *E. coli* K12 was mutagenized with *N*-methyl-*N*′-nitro-*N*-nitrosoguanidine and a replica plating technique was used to select mutants with a temperature-dependent intolerance for oxygen. One of these mutants exhibited a temperature-dependent decline in its content of SOD, which paralleled its intolerance for oxygen.[71] In a more recent study several oxygen-intolerant mutants of *E. coli* K12 were studied. One group of these had lost the ability to induce catalase and peroxidase in response to oxygen exposure, while the second group was also unable to induce Mn SOD. When members of the first group were prevented from inducing Mn SOD by puromycin, their oxygen-intolerance was markedly augmented.[72] This behavior demonstrates the importance of scavenging both $O_2^{\cdot -}$ and H_2O_2 and also suggests some genetic linkage between hydroperoxidases and the Mn SOD in *E. coli.* Further work with mutants is needed.

XIII. SUMMARY

These data lead to the conclusion that $O_2^{\cdot -}$ can be generated within respiring cells and would seriously damage those cells, were it not for SODs which, by scavenging this radical, keep its steady-state concentration vanishingly small. In the case of *L. plantarum,* and possibly in other cases not yet discovered, the role of $O_2^{\cdot -}$-scavenger is played by millimolar levels of Mn(II).[21] Increasing the rate of intracellular $O_2^{\cdot -}$ production, whether by raising the pO_2, increasing the growth rate in chemostat culture, replacing glucose by nonfermentable carbon sources, adding redox active compounds capable of increasing $O_2^{\cdot -}$ production, or by inhibiting cytochrome oxidase, leads to an adaptive increase in the synthesis of SOD. Blocking this adaptive response with inhibitors of protein synthesis such as puromycin or chloramphenicol, or with a nutritionally restricted medium, causes the cells to succumb to levels of $O_2^{\cdot -}$ production which they would otherwise survive.

In *E. coli* the constitutive Fe SOD provides a back-up defense always ready in the event of sudden exposure to O_2, while the Mn SOD, which is under repression control, allows fine tuning of the level of defense to the level of threat. In the case of some obligate anaerobes, an Fe SOD seems to serve the same purpose as it does in anaerobically grown *E. coli,* i.e., assuring survival during transient exposure to oxygen.

We have not discussed H_2O_2 and the enzymes which catalytically scavenge it. This is not to be taken as an indication that they are not important, but rather that a book devoted to superoxide is not an appropriate place in which to summarize data bearing on H_2O_2, catalases, and peroxidases. A balanced view would, however, recognize that $O_2^{\cdot -}$ and H_2O_2 are both important agents of oxygen toxicity and that prospering in the presence of O_2 requires enzymic defenses against both of these intermediates of oxygen reduction.

REFERENCES

1. **McCord, J. M. and Fridovich, I.**, The reduction of cytochrome *c* by milk xanthine oxidase, *J. Biol. Chem.*, 243, 5753, 1968.
2. **McCord, J. M. and Fridovich, I.**, Superoxide dismutase—an enzymic function for erythrocuprein, *Fed. Proc.*, 28, 346, 1968.
3. **McCord, J. M. and Fridovich, I.**, Superoxide dismutase—an enzymic function of erythrocuprein (hemocuprein), *J. Biol. Chem.*, 244, 6049, 1969.
4. **McCord, J. M., Keele, B. B., Jr., and Fridovich, I.**, An enzyme-based theory of obligate anaerobiosis: the physiological function of superoxide dismutase, *Proc. Natl. Acad. Sci. U.S.A.*, 68, 1024, 1971.
5. **Lindmark, D. G. and Müller, M.**, Superoxide dismutase in the anaerobic flagellates, *Tritrichomonas foetus* and *monocercomonas* sp., *J. Biol. Chem.*, 249, 4634, 1974.
6. **Hewitt, J. and Morris, J. G.**, Superoxide dismutase in some obligately anaerobic bacteria, *FEBS Lett.*, 50, 315, 1975.
7. **Morris, J. G.**, Fifth Stenhouse-Williams memorial lecture. Oxygen and the obligate anaerobe, *J. Appl. Bacteriol.*, 40, 229, 1976.
8. **Gregory, E. M., Kowalski, J. B., and Holdeman, L. V.**, Production and some properties of catalase and superoxide dismutase from the anaerobe *Bacteroides distasonis, J. Bacteriol.*, 129, 1298, 1977.
9. **Hatchikian, E. C. and Henry, Y. A.**, An iron-containing superoxide dismutase from the strict anaerobic *Desulfovibrio desulfuricans* (Norway 4), *Biochimie*, 59, 153, 1977.
10. **Tally, F. P., Goldin, B. R., Jacobus, N. V., and Gorbach, S. L.**, Superoxide dismutase in anaerobic bacteria of clinical significance, *Infect. Immunol.*, 16, 20, 1977.
11. **Gregory, E. M., Moore, W. E. C., and Holdeman, L. V.**, Superoxide dismutase in anaerobes: survey, *Appl. Environ. Microbiol.*, 35, 988, 1978.
12. **Uesugi, I. and Yajima, M.**, Oxygen and strictly anaerobic intestinal bacteria. II. Oxygen metabolism in strictly anaerobic bacteria, *Z. Allg. Microbiol.*, 18, 593, 1978.
13. **Kanematsu, S. and Asada, K.**, Superoxide dismutase from an anaerobic photosynthetic bacterium, *Chromatium vinosum, Arch. Biochem. Biophys.*, 185, 473, 1978.
14. **Wimpenny, J. W. T. and Samah, O. A.**, Some effects of oxygen on the growth and physiology of *Selenomonas ruminantium, J. Gen. Microbiol.*, 108, 329, 1978.
15. **Gregory, E. M.**, Chemical and physical differentiation of superoxide dismutases in anaerobes, *Fed. Proc.*, 39, 1773, 1980.
16. **Levitt, M. D. and Bond, J. H., Jr.**, Volume, composition, and source of intestinal gas, *Gastroenterology*, 59, 921, 1970.
17. **Privalle, C. T. and Gregory, E. M.**, Superoxide dismutase and O_2 lethality in *Bacteroides fragilis J. Bacteriol.*, 138, 139, 1979.
18. **Gregory, E. M. and Fridovich, I.**, Oxygen metabolism in *Lactobacillus plantarum, J. Bacteriol.*, 117, 166, 1974.
19. **Strittmatter, C. F.**, Electron transport to oxygen in *Lactobacilli.* And, Flavin-linked oxidative enzymes of *Lactobacillus casei, J. Biol. Chem.*, 234, 2789 and 2794, 1959.
20. **Yousten, A. A., Johnson, J. L., and Salin, M.**, Oxygen metabolism of catalase-negative and catalase-positive strains of *Lactobacillus plantarum, J. Bacteriol.*, 123, 242, 1975.
21. **Archibald, F. S. and Fridovich, I.**, Defenses against oxygen toxicity in *Lactobacillus plantarum, J. Bacteriol.*, 145, 442, 1981.
22. **Gregory, E. M. and Fridovich, I.**, The induction of superoxide dismutase by molecular oxygen, *J. Bacteriol.*, 114, 543, 1973.
23. **Gregory, E. M. and Fridovich, I.**, Oxygen toxicity and the superoxide dismutase, *J. Bacteriol.*, 114, 1193, 1973.
24. **Gregory, E. M., Yost, F. J., and Fridovich, I.**, The superoxide dismutase of *Escherichia coli*: intracellular localization and functions, *J. Bacteriol.*, 115, 987, 1973.
25. **Crapo, J. D. and Tierney, D.**, Superoxide dismutase and oxygen toxicity, *Clin. Res.*, 21, 222, 1973.
26. **Crapo, J. D. and Tierney, D.**, Superoxide dismutase and pulmonary oxygen toxicity, *Am. J. Physiol.*, 226, 1401, 1974.
27. **Kimball, R. E., Reddy, K., Pierce, T. H., Schwartz, L. W., Mustafa, M. G., and Cross, C. E.**, Oxygen toxicity: augmentation of antioxidant defense mechanisms in rat lung, *Am. J. Physiol.*, 230, 1425, 1976.
28. **Autor, A. P., Frank, L., and Roberts, R. J.**, Developmental characteristics of pulmonary superoxide dismutase: relationship to idiopathic respiratory distress syndrome, *Pediatr. Res.*, 10, 154, 1976.
29. **Stevens, J. B. and Autor, A. P.**, Induction of superoxide dismutase by oxygen in neonatal rat lung, *J. Biol. Chem.*, 252, 3509, 1977.
30. **Autor, A. P. and Stevens, J. B.**, Mechanism of oxygen detoxification in neonatal rat lung tissue, *Photochem. Photobiol.*, 28, 775, 1978.

31. **Roberts, R. J.**, Employment of pulmonary superoxide dismutase, catalase, and glutathione peroxidase activity as criteria for assessing suitable animal models for studies of bronchopulmonary dysplasia, *J. Pediatr.*, 95, 904, 1979.
32. **Gregory, E. M., Goscin, S. A., and Fridovich, I.**, Superoxide dismutase and oxygen toxicity in a eukaryote, *J. Bacteriol.*, 117, 456, 1974.
33. **Puget, K. and Michelson, A. M.**, Isolation of a new copper-containing superoxide dismutase bacteriocuprein, *Biochem. Biophys. Res. Commun.*, 58, 830, 1974.
34. **Rister, M. and Baehner, R. L.**, Induction of superoxide dismutase (SOD) activity *in vivo* by oxygen (O_2) in polymorphonuclear leukocytes (PMNS) and alveolar macrophages (AM), *Blood*, 46, 1016, 1975.
35. **Rister, M. and Baehner, R. L.**, The alteration of superoxide dismutase, catalase, glutathione peroxidase, and NAD(P)H cytochrome *c* reductase in guinea pig polymorphonuclear leukocytes and alveolar macrophages during hyperoxia, *J. Clin. Invest.*, 58, 1174, 1976.
36. **Stevens, J. B. and Autor, A. P.**, Oxygen-induced synthesis of superoxide dismutase and catalase in pulmonary macrophages of neonatal rats, *Lab. Invest.*, 37, 470, 1977.
37. **Nerurkar, L. S., Zeligs, B. J., and Bellanti, J. A.**, Changes in superoxide dismutase, catalase and glucose-6-phosphate dehydrogenase activities of rabbit alveolar macrophages; induced by post-natal maturation and/or *in vitro* hyperoxia, *Photochem. Photobiol.*, 28, 781, 1978.
38. **Petkau, A., Monasterski, L. G., Kelly, K., and Friesen, H. G.**, Modification of superoxide dismutase in rat mammary carcinoma, *Res. Commun. Chem. Pathol. Pharmacol.*, 17, 125, 1977.
39. **Pritchard, G. G., Wimpenny, J. W. T., Morris, H. A., Lewis, M. W. A., and Hughes, D. E.**, Effects of oxygen on *Propionbacterium shermanii* grown in continuous culture, *J. Gen. Microbiol.*, 102, 223, 1977.
40. **Boveris, A., Sánchez, R. A., and Beconi, M. T.**, Antimycin- and cyanide-resistant respiration and superoxide anion production in fresh and aged potato tuber mitochondria, *FEBS Lett.*, 92, 333, 1978.
41. **Ghosh, S. and Chatterjee, G. C.**, Superoxide dismutase activity in *Vibrio eltor* in relation to oxygen toxicity and the bactericidal action of nitrofurantoin, *J. Gen. Appl. Microbiol.*, 25, 367, 1979.
42. **Friedberg, D., Fine, M., and Oren, A.**, Effect of oxygen on the cyanobacterium *Oscillatoria limnetica*, *Arch. Microbiol.*, 123, 311, 1979.
43. **Hassan, H. M. and Fridovich, I.**, Enzymatic defenses against the toxicity of oxygen and of streptonigrin in *Escherichia coli* K12, *J. Bacteriol.*, 129, 1574, 1977.
44. **Hassan, H. M. and Fridovich, I.**, The physiological function of superoxide dismutase in *Escherichia coli* K12: studies of glucose-limit ed chemostat cultures, *J. Bacteriol.*, 130, 805, 1977.
45. **Hassan, H. M. and Fridovich, I.**, Regulation of the synthesis of superoxide dismutase in *E. coli*: the glucose effect, *J. Bacteriol.*, 132, 505, 1977.
46. **Hassan, H. M. and Fridovich, I.**, Regulation of the synthesis of superoxide dismutase in *E. coli*: induction by methyl viologen, *J. Biol. Chem.*, 252, 7667, 1977.
47. **Hassan, H. M. and Fridovich, I.**, Superoxide radicals and the oxygen enhancement of the toxicity of paraquat in *Escherichia coli*, *J. Biol. Chem.*, 253, 8143, 1978.
48. **Hassan, H. M. and Fridovich, I.**, Intracellular production of superoxide radical and of hydrogen peroxide by redox active compounds, *Arch. Biochem. Biophys.*, 196, 385, 1979.
49. **Hassan, H. M. and Fridovich, I.**, Mechanism of the antibiotic action of pyocyanine, *J. Bacteriol.*, 141, 156, 1980.
50. **Boveris, A. and Chance, B.**, The mitochondrial generation of hydrogen peroxide, *Biochem. J.*, 134, 707, 1973.
51. **Forman, H. J. and Kennedy, J. A.**, Role of superoxide radical in mitochondrial dehydrogenase reactions, *Biochem. Biophys. Res. Commun.*, 60, 1044, 1974.
52. **Loschen, G., Azzi, A., Richler, C., and Flohé, L.**, Superoxide radicals as precursors of mitochondrial hydrogen peroxide, *FEBS Lett.*, 42, 68, 1974.
53. **Tyler, D. D.**, A protective function of superoxide dismutase during respiratory chain activity, *Biochim. Biophys. Acta*, 396, 335, 1975.
54. **Boveris, A. and Cadenas, E.**, Mitochondrial production of superoxide anions and its relationship to the antimycin insensitive respiration, *FEBS Lett.*, 54, 311, 1975.
55. **Boveris, A.**, Mitochondrial production of superoxide radical and hydrogen peroxide, *Adv. Exp. Med. Biol.*, 78, 67, 1977.
56. **Takeshige, K. and Minakami, S.**, NADH and NADPH-dependent formation of superoxide anions by bovine heart submitochondrial particles and NADH-ubiquinone reductase preparations, *Biochem. J.*, 180, 129, 1979.
57. **Farrington, J. A., Ebert, M., Land, E. J., and Fletcher, K.**, Bipyridylium quaternary salts and related compounds. V. Pulse radiolysis studies of the reaction of paraquat radical with oxygen. Implications for the mode of action of bipyridyl herbicides, *Biochim. Biophys. Acta*, 314, 372, 1973.
58. **Hassan, H. M. and Fridovich, I.**, Paraquat and *Escherichia coli*: mechanism of production of extracellular superoxide radical, *J. Biol. Chem.*, 254, 10846, 1979.

59. **Babior, B., Curnutte, J. T., and Kipnes, R. S.,** Biological defense mechanisms. Evidence for the participation of superoxide in bacterial killing by xanthine oxidase, *J. Lab. Clin. Med.,* 85, 235, 1975.
60. **Ismail, G., Sawyer, W. D., and Wegener, W. S.,** Effect of hydrogen peroxide and superoxide radical on viability of *Neisseria gonorrhoeae* and related bacteria, *Proc. Soc. Exp. Biol. Med.,* 155, 264, 1977.
61. **Kellogg, E. W., III, Yost, M. G., Barthakur, N., and Kreuger, A. P.,** Superoxide involvement in the bactericidal effects of negative air ions on *Staphylococcus albus, Nature (London),* 281, 400, 1979.
62. **Hoffman, P. S., George, H. A., Krieg, N. R., and Smibert, R. M.,** Studies of the microaerophilic nature of Campylobacter fetus subsp. *jejuni.* II. Role of exogenous superoxide anions and hydrogen peroxide, *Can. J. Microbiol.,* 25, 8, 1979.
63. **Rosen, H. and Klebanoff, S. J.,** Bactericidal activity of a superoxide anion-generating system. A model for the polymorphonuclear leukocyte, *J. Exp. Med.,* 149, 27, 1979.
64. **Lavelle, F., Michelson, A. M., and Dimitrijevic, L.,** Biological protection by superoxide dismutase, *Biochem. Biophys. Res. Commun.,* 55, 350, 1973.
65. **Michelson, A. M. and Buckingham, M. E.,** Effects of superoxide radicals on myoblast growth and differentiation, *Biochem. Biophys. Res. Commun.,* 58, 1079, 1974.
66. **Petkau, A., Chelack, W. S., and Pleskach, S. D.,** Protection of post-irradiated mice by superoxide dismutase, *Int. J. Radiat. Biol.,* 29, 297, 1976.
67. **Misra, H. P. and Fridovich, I.,** Superoxide dismutase and the oxygen enhancement of radiation lethality, *Arch. Biochem. Biophys.,* 176, 577, 1976.
68. **Oberley, L. W., Lindgren, A. L., Baker, S. A., and Stevens, R. H.,** Superoxide ion as the cause of the oxygen effect, *Radiat. Res.,* 68, 320, 1976.
69. **Niwa, T., Yamaguchi, H., and Yano, K.,** Radioprotection by superoxide dismutase: reduction of oxygen effect, in *Biochemical and Medical Aspects of Active Oxygen,* Hayaishi, O. and Asada, K., Eds., University of Tokyo Press, Tokyo, 1977, 209.
70. **Samuni, A., Chevion, M., Halpern, Y. S., Ilan, Y. A., and Czapski, G.,** Radiation-induced damage in T4 bacteriophage: the effect of superoxide radicals and molecular oxygen, *Radiat. Res.,* 75, 489, 1978.
71. **McCord, J. M., Beauchamp, C. O., Goscin, S., Misra, H. P., and Fridovich, I.,** Superoxide and superoxide dismutase, in *Oxidases and Related Redox Systems,* Proc. 2nd Int. Symp., King, T. E., Mason, H. S., and Morrison, M., Eds., University Park Press, Baltimore, 1973, 66.
72. **Hassan, H. M. and Fridovich, I.,** Superoxide, hydrogen peroxide and oxygen tolerance of oxygen-sensitive mutants of *Escherichia coli, Rev. Infect. Dis.,* 1, 357, 1979.

Chapter 5

THE TOXIC EFFECTS OF OXYGEN ON PLANT TISSUES

Barry Halliwell

TABLE OF CONTENTS

I. TOXIC EFFECTS OF OXYGEN ON PLANT TISSUES

Oxygen is, by definition, essential for the life of all aerobic organisms, including higher plants, but it has long been known to be toxic to them at concentrations higher than those in normal air. High oxygen concentrations have been found to inhibit chloroplast development,[1] decrease seed viability and root growth,[4] damage the membranes of leaves and roots, stimulate leaf abscission,[2] and increase the incidence of growth abnormalities.[3] Exposure to oxygen at 6 atm pressure for 15 hr was lethal to a wide variety of plants, although observable damage, in the form of desiccation, shriveling and discoloration of leaves, and the collapse of soft stems, only became apparent after the end of the treatment period.[4]

When soybean plants were grown at low (5%) oxygen concentrations, the growth of leaves, stems, and roots was actually increased;[207] this suggests that even 21% O_2 is sufficient to produce a growth-inhibiting effect. However, growth at low O_2 hindered seed development in the plants, and so some part of the reproductive process must require O_2 concentrations higher than 5%. The nature of this remains to be established.[207]

Oxygen toxicity is seen not only in whole plants but also in cell cultures and even in isolated organelles. For example, high O_2 concentrations inhibit CO_2 fixation by isolated intact chloroplast fractions (the so-called "Warburg effect"). Chloroplasts are especially prone to oxygen-toxicity effects because their internal O_2 concentration during photosynthesis is always somewhat greater than that in the air surrounding the leaf.[5]

The chloroplasts present in the leaves of higher plants can be seen under the electron microscope to be bounded by an outer envelope consisting of two membranes separated by an electron-translucent space of about 10 nm (Figure 1). The envelope encloses the "stroma" of the chloroplast, in which floats a complex internal membrane structure. The stroma is an aqueous solution containing various low-molecular-weight compounds plus a high concentration of proteins, most of which are the enzymes necessary to convert CO_2 into carbohydrate by a complex metabolic pathway known as the "Calvin cycle". This pathway is described in Figure 2.

The internal membrane structure of the chloroplast, which floats in the stroma, is extremely complex. In the electron micrograph shown in Figure 1, two distinct features may be recognized, viz., regions of closely stacked membranes known as "grana" that are interconnected by a three-dimensional network of membranes known as the "stroma thylakoids". The membranes of this complex internal structure contain the photosynthetic pigments (mainly chlorophylls *a* and *b*, green pigments that absorb light in the blue and red regions of the spectrum, together with carotenoids and xanthophylls, yellow pigments absorbing blue light) and they function to produce the reduced form of nicotinamide adenine dinucleotide phosphate (NADPH) and, by the process of "photophosphorylation", the adenosine 5′-triphosphate (ATP) needed to drive CO_2 fixation by the Calvin cycle in the stroma (Figure 2).

Absorption of light energy by chlorophyll or other pigment molecules results in the formation of higher electronic excitation states (excited singlet states). Absorption of a quantum of blue light by chlorophyll results in formation of a more excited singlet state (second excited state) than does absorption of a quantum of red light (Figure 3), but the second excited state loses its excess energy as heat extremely quickly, so that absorption of either red or blue quanta effectively produces the same first excited state of the chlorophyll molecules.[6] Energy so absorbed can be lost by reemission of a photon to give fluorescence; it can also be lost as heat or it can be transferred to another molecule, e.g., an adjacent chlorophyll, the first molecule returning to the ground state while the second becomes excited. Energy absorbed by carotenoids can be directly transferred to chlorophyll molecules in this way.[6,7]

FIGURE 1. Structure of the chloroplasts in higher plant cells as seen under the electron microscope. (Mesophyll chloroplasts are shown at a magnification of 12,500 ×.)

Certain special molecules of chlorophyll *a* in the thylakoid membrane (reaction center chlorophylls) can, when excited, lose an electron to a neighboring electron acceptor (A), producing a chlorophyll$^+$ A$^-$ pair. This initial charge separation is the basic reaction of photosynthesis. Light energy absorbed by other chlorophyll and carotenoid molecules can be transferred quite efficiently in the thylakoid membrane to the reaction center chlorophylls. Indeed, each reaction center chlorophyll is associated with a "light-harvesting array" of other pigment molecules that can funnel energy to it.

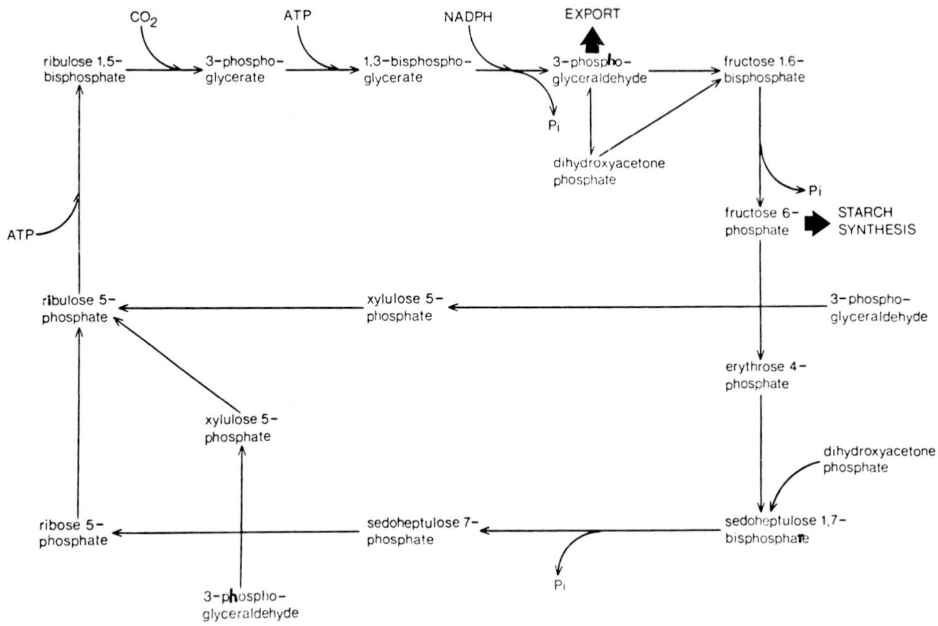

FIGURE 2. Reactions of the Calvin cycle. CO_2 is fixed by the enzyme ribulose diphosphate carboxylase, which produces two molecules of phosphoglycerate (2 × C_3 compounds). Five of the carbon atoms must be recycled to replace the acceptor ribulose diphosphate molecule. For every three CO_2 molecules fixed, an extra C_3 compound is generated which can be exported from the chloroplast for metabolism in the leaf cytoplasm, converted to hexose phosphates for starch synthesis in the chloroplast, or used to increase the concentration of Calvin cycle intermediates.

Two different types of reaction center, known as P700 and P680, can be detected in the chloroplasts of higher plants.[7] Each is served by its own light-harvesting pigment antenna system: the complexes containing P700 are referred to as photosystems I and those containing P680 as photosystems II. Electrons ejected from the P680 reaction center of photosystem II are accepted by a molecule of unknown structure, usually referred to as "Q". From Q the electrons pass "down" (i.e., to less negative redox potentials) an electron transport chain that contains plastoquinone, plastocyanin, and cytochrome *f* and they eventually enter photosystem I to replace the electrons lost on excitation of the P700 reaction center (Figure 4). The molecule which accepts excited electrons from photosystem I has not been conclusively identified, but there is evidence that it may be a non-haem-iron-sulfur protein. From this primary acceptor electrons pass to other similar proteins bound to the membranes and then onto a soluble (stromal) iron-sulfur protein, known as ferredoxin. Soluble ferredoxin from higher plants contains two atoms of iron and two of sulfur per molecule and it acts as a one-electron acceptor. Reduced ferredoxin can then, in the presence of a reductase enzyme, donate electrons to $NADP^+$ to give NADPH (Figure 4). The electrons ejected from the photosystem II reaction centers are replaced by the splitting of water molecules, accompanied by O_2 evolution.

$$2H_2O \longrightarrow 4e^- + 4H^+ + O_2 \qquad (1)$$

Manganese and chloride ions are required for the action of photosystem II. Further details of all these processes may be found in recent reviews.[6-9] Oxygen can damage plant tissues by several mechanisms, which are considered below.

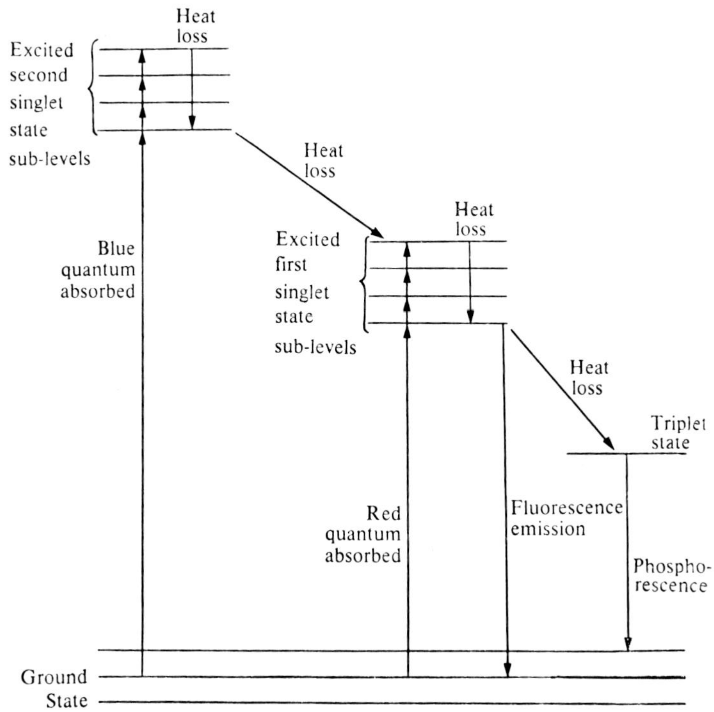

FIGURE 3. Energy changes in the chlorophyll molecule. Absorption of light energy by the chlorophyll molecule results in the formation of higher electronic excitation states. Although the energy stored by the second singlet state is much greater than that in the first singlet state, the excess is lost as heat and does not contribute to photosynthesis. Both ground and excited states possess a series of subenergy levels. Excitation can proceed from any sublevel of the ground state to any sublevel of an excited state, but heat loss occurs to give the lowest subenergy level of the excited state. Hence any energy emitted by fluorescence is usually less than that absorbed during excitation and so the fluorescence spectrum is shifted to the red end of the spectrum as compared with the spectrum of the exciting light (Stokes shift). A triplet state can also be formed from the excited singlet states. This arises because an orbital can contain a maximum of two electrons, each of which much have a different value of the spin quantum number ($+\frac{1}{2}$ or $-\frac{1}{2}$). When one of the electrons is promoted to a higher orbital by absorption of energy or by other means, the value of its spin quantum number is not normally changed. However, spin reversal can take place infrequently during electron promotion, and the promoted electron and the remaining electron in the orbital in the ground state both have the same spin quantum number. Direct light emission from the triplet state (phosphorescence) has to be accompanied by spin reversal and has a longer wavelength than normal fluorescence. Alternatively, the triplet can be converted back to the singlet, a process with low probability. Nevertheless, this requires only a small amount of energy and can result in a "delayed" fluorescence with the normal fluorescence spectrum. Triplet chlorophyll states are not thought to play a role in photosynthesis, although their formation has been observed in vivo. (From Gregory, R. P. F., *Biochemistry of Photosynthesis,* 2nd ed., John Wiley & Sons, London, 1977. With permission.)

A. Diversion of Normal Metabolic Pathways into Abnormal Routes

The first reaction of the Calvin cycle is the carboxylation of ribulose diphosphate to give two molecules of phosphoglyceric acid, catalyzed by the stromal enzyme ribulose diphosphate carboxylase (Figure 2). This enzyme has very complex properties, which have been discussed elsewhere.[8,9] In particular, one of the chemical intermediates formed during the action of the enzyme on ribulose diphosphate is capable of

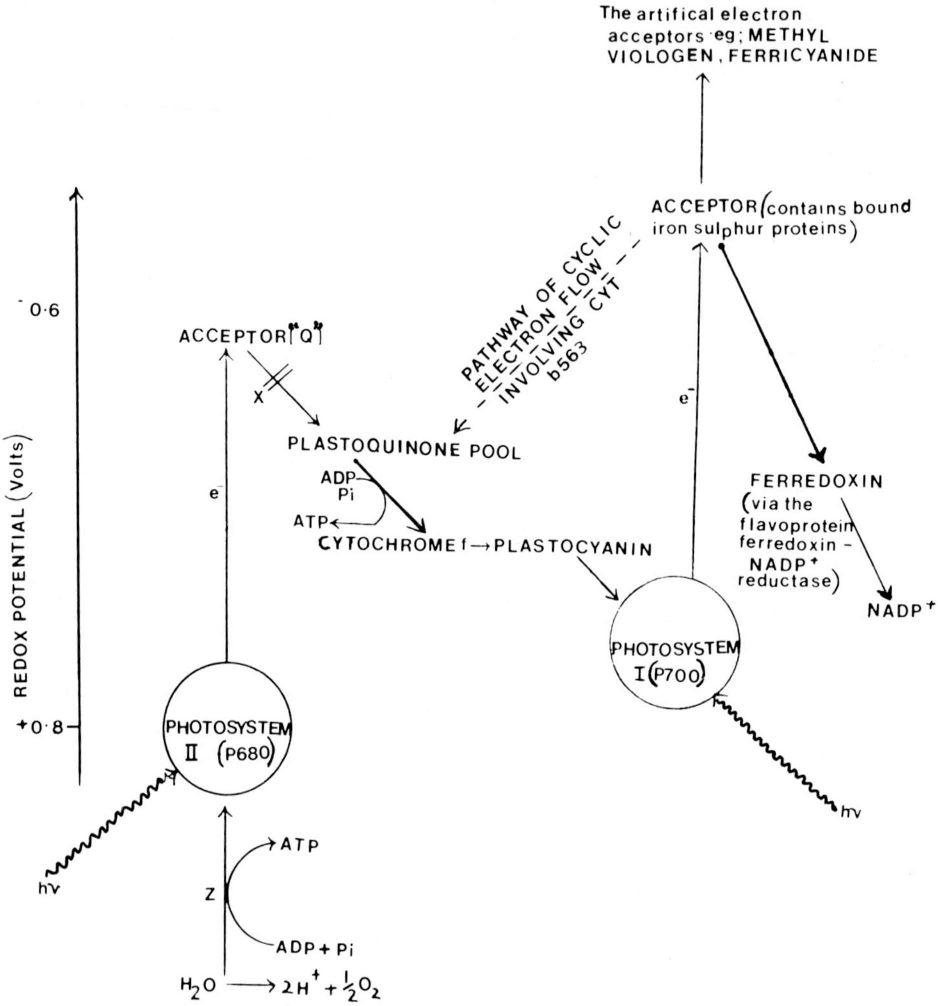

FIGURE 4. Current scheme of photosynthetic electron transport. This is often known as the Hill-Bendall scheme or the "Z" scheme. The sequence of carriers is presented as an energy diagram using redox potentials. The lower the redox potential of a substance, the better is its electron-donating capacity. Hence a component with a negative redox potential is theoretically capable of donating electrons to another component with a less negative, zero, or positive redox potential. The redox potentials of the components are obtained by titrating them with reagents of known redox potential. X represents the site of action of the commonly used photosynthetic inhibitor, DCMU.

reacting with molecular oxygen as shown in Figure 5. As a result of this "oxygenase" activity of the enzyme, ribulose diphosphate becomes converted into phosphoglyceric acid and phosphoglycollic acid. The phosphoglycollate so produced is further metabolized by a series of reactions known as the "photorespiratory pathway". Figure 6 shows this pathway, which essentially converts two molecules of phosphoglycollate into CO_2 and phosphoglycerate for reentry into the Calvin cycle.[8,9] Hence CO_2 previously fixed into the Calvin cycle intermediates is lost again and has to be refixed with expenditure of more energy. Both O_2 and CO_2 compete for the enzyme intermediate (Figure 5). The affinity of the carboxylase for CO_2 is much greater than that for O_2, but at elevated O_2 concentrations CO_2 fixation into the Calvin cycle is decreased and CO_2 loss by photorespiration is increased. At some O_2/CO_2 ratios, a point will be reached at which the leaf is no longer achieving any net CO_2 fixation, and therefore

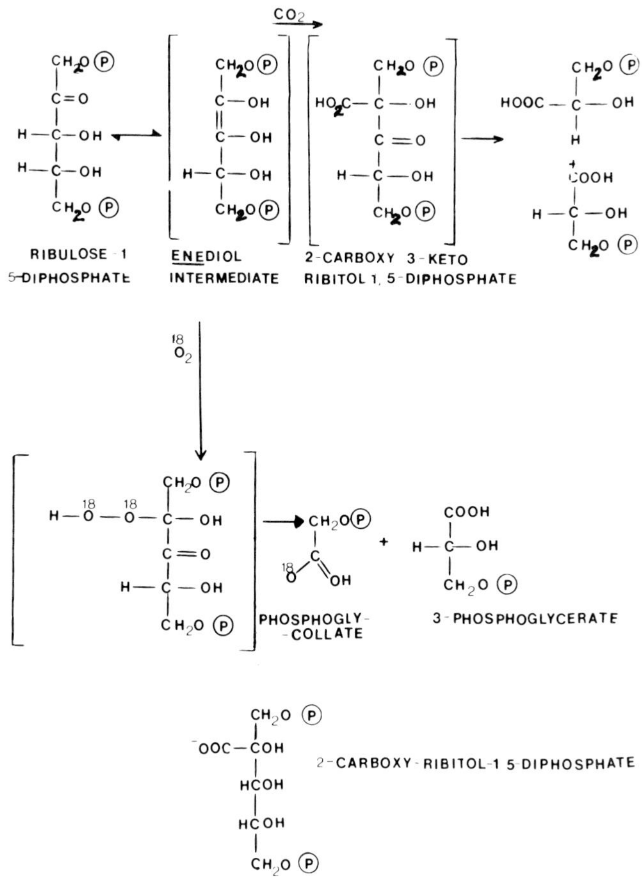

FIGURE 5. Proposed mechanism for carboxylase and oxygenase activities of ribulose diphosphate carboxylase.

plant growth is completely inhibited. At high O_2/CO_2 ratios there may be actual loss of carbon from the plant as photorespiration exceeds CO_2 fixation into the Calvin cycle, a situation which will result in death of the plant if continued for a long period. The oxygenase activity of ribulose diphosphate carboxylase accounts for a large part of the damaging effects of elevated O_2 on leaves and on isolated chloroplasts. It is not the whole story, however, even for these systems,[8,9] and it obviously cannot account for the toxic effects of oxygen on roots and other nongreen plant tissues. We must therefore look at alternative mechanisms.

B. Inactivation of Enzymes

This was one of the earliest proposals made to explain O_2 toxicity.[10] Enzyme inactivation is best illustrated by the case of certain anaerobes, which stop growing or die immediately on exposure to O_2.[11] For example, the activity of the nitrogenase enzyme of the anaerobic bacterium *Clostridium pasteurianum* depends on maintaining some of its co-factors in a highly reduced state. On exposure to O_2 they are irreversibly oxidized and the enzyme is inactivated.[12] The nitrogen-fixing enzymes of blue-green algae are also oxygen sensitive, which poses a problem for unicellular algae such as *Gloeocapsa* that carry out nitrogen fixation and photosynthesis in the same cell.[13] In order to prevent inactivation of the nitrogenase, the period of maximum nitrogenase activity occurs at a time in the life cycle when the rate of photosynthesis is low. The

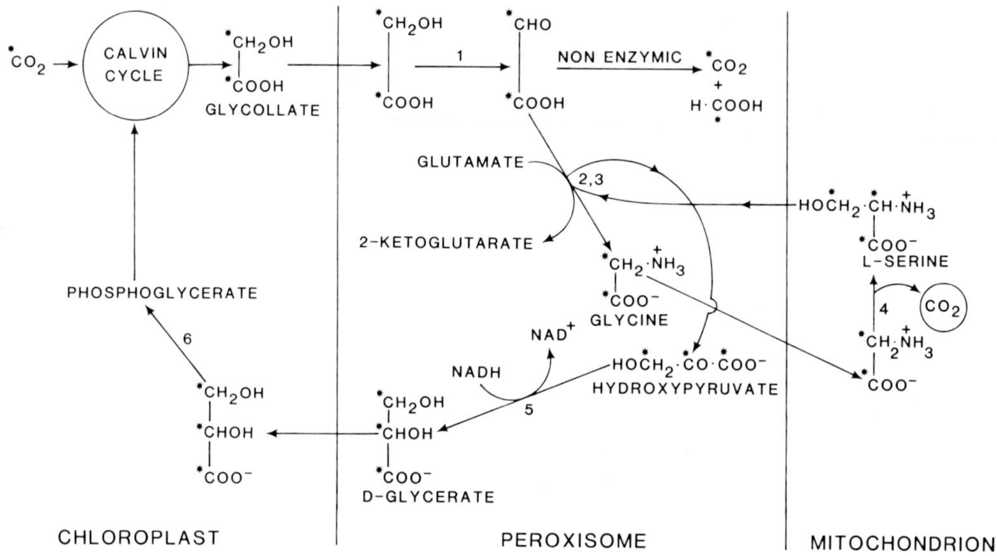

FIGURE 6. Scheme for CO_2 release during photorespiration. Phosphoglycollate is hydrolyzed to glycollate by a phosphatase in the chloroplast stroma. The glycollate so produced leaves the chloroplast and is oxidized to glyoxylate in leaf peroxisomes. Some glyoxylate is converted to CO_2 and formate in a nonenzymic reaction, but most is transaminated to yield glycine. Glycine moves from peroxisomes to mitochondria, where two molecules of glycine form one molecule of serine and CO_2 is released. The C_3 compound serine eventually reforms phosphoglycerate, which reenters the Calvin cycle. For every two molecules of glycollate lost from the chloroplast (four carbon atoms), one phosphoglycerate is produced (three carbon atoms) and so photorespiration results in a net loss of carbon from the chloroplast. If not all the phosphoglycerate reenters the chloroplast (e.g., it has been suggested that some is used to make sucrose in the cytoplasm) the net loss of carbon is greater. Enzymes: (1) glycollate oxidase; (2) glutamate-glyoxylate aminotransferase; (3) serine-glyoxylate aminotransferase; (4) glycine decarboxylase and serine hydroxymethyltransferase; (5) D-glycerate dehydrogenase; (6) phosphoglycerate kinase.

cells are also well provided with the various protective mechanisms described later in this chapter.[13] In other blue-green algae, such as *Anabaena cylindrica*, nitrogen fixation occurs in specialized cells (heterocysts) that lack an O_2-evolving mechanism.[209]

Some enzymes present in aerobes are also inactivated by O_2, although much less rapidly than enzymes such as nitrogenase. Inactivation is usually due to the oxidation of essential thiol (−SH) groups on the enzymes. A particularly oxygen-sensitive Calvin cycle enzyme is $NADP^+$-glyceraldehyde-3-phosphate dehydrogenase, which becomes inhibited when chloroplasts are exposed to high oxygen tensions.[14] Fructose diphosphatase (Figure 2) is also easily oxidized into a form that cannot participate in the Calvin cycle.[15] Other oxygen-sensitive enzymes in some plants include pyruvate-phosphate dikinase.[16]

Plant cells have evolved a mechanism to prevent inactivation of enzymes by oxidation of thiol groups; the cells contain high (millimolar) concentrations of a low-molecular-weight thiol compound, the simple tripeptide glutathione (Figure 7). Since reduced glutathione (GSH) is more easily "available" to O_2 than are enzyme thiol groups, it should be preferentially oxidized and will therefore protect the enzymes. GSH can also reactivate some enzymes that have been inhibited by previous exposure to high O_2 concentrations, presumably by reducing their oxidized −SH groups.

Reaction of GSH with O_2 gives the oxidized form (GSSG), which contains a disulfide bridge (Figure 7).

$$2GSH + \tfrac{1}{2}O_2 \longrightarrow GSSG + H_2O \qquad (2)$$

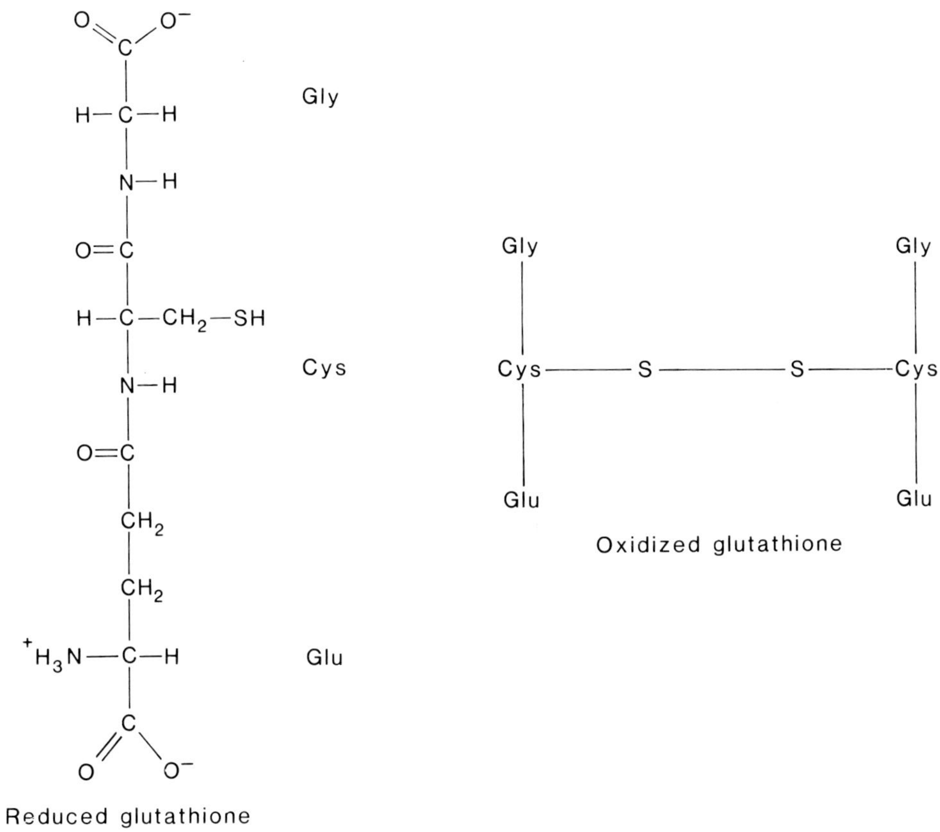

FIGURE 7. Structure of glutathione (γ-L-glutamyl-L-cysteinylglycine).

The oxidation of pure GSH by O_2 is very slow, but it is catalyzed by traces of metal ions and other co-factors and proceeds rapidly in extracts of plant tissues,[156,162] especially at alkaline pH values. The chloroplast stroma contains glutathione at millimolar concentrations[17,18] and becomes alkaline during illumination, because the operation of the electron-transport chain causes protons to be translocated from the stroma into the thylakoids to set up a proton gradient that can be used to drive ATP synthesis by the chemiosmotic mechanism.[6,8,9] Similar concentrations of glutathione are found in other plant tissues.[19,156,157]

GSSG may be converted back to GSH by the activity of glutathione reductase enzymes, which are usually NADPH dependent (Equation 3) Their equilibrium position greatly favors

$$GSSG + NADPH + H^+ \longrightarrow 2GSH + NADP^+ \qquad (3)$$

reduction of GSSG.[19] Both the chloroplast and the cytoplasm of leaf tissues contain an NADPH-dependent glutathione reductase,[18,20,21] and this enzyme has been reported in many other plant tissues as well.[19,22,158-160] The purified spinach chloroplast enzyme has a molecular weight of about 145,000 and consists of two apparently identical subunits. It has a broad pH optimum (7.8 to 9.3). Its K_m for GSSG is approximately 200 μM and that for NADPH is approximately 3 μM. The concentration of NADPH in chloroplasts in the dark has been given[23] as 15 nmol/mg chlorophyll, and this concentration is higher in the light because of the operation of the electron transport chain (Figure 4). If we assume an "available internal volume" of 21 $\mu\ell$/mg chlorophyll[24] for spinach chloroplasts, their internal concentration of NADPH in the dark may be calculated as 710 μM, which is amply sufficient to saturate the glutathione reductase. The enzyme will therefore be half-maximally active at a GSSG concentration of 200 μM, which is small in relation to the total glutathione content of chloroplasts (1.0 to 3.5 mM). Since substantial amounts of glutathione reductase activity are present in chloroplasts, one would expect stromal GSH/GSSG ratios to be normally kept high under both light and dark conditions, a prediction which has been confirmed experimentally.[21] A high ratio of GSH/GSSG has also been reported in other plant tissues.[19,25,162]

As plant tissues or isolated chloroplasts are exposed to increasing O_2 concentrations, however, a point might be reached at which Reaction 2 proceeds more rapidly than Reaction 3 and GSH/GSSG ratios begin to fall, which would allow inactivation of enzymes such as $NADP^+$-glyceraldehyde-3-phosphate dehydrogenase.[14,18] GSSG itself can inactivate enzymes by forming mixed disulfides with them, as shown in Equation 4.

$$\text{Enzyme} - \text{SH} + \begin{matrix} \text{GS—S} \\ | \\ \text{GS—S} \end{matrix} \rightleftharpoons \text{enzyme} - \text{S} - \text{SG} + \text{GSH}$$

Active enzyme Inactive enzyme (4)

Mixed disulfides have been found in vivo in both bacterial and in animal systems[10,26-29,60] and should presumably be formed in plant cells as well.[162] The occurrence of Reaction 4 provides a good reason for plant tissues to keep their GSSG concentrations low under normal conditions.

GSH also plays a role in removing hydrogen peroxide in chloroplasts, as discussed later in this chapter. In some plant tissues it is involved in the metabolism of phenolic compounds,[30] as an enzyme co-factor in the oxidation of formaldehyde into formic acid[161] and in the detoxification of herbicides such as atrazine, fluorodifen, propachlor, barban, and CDAA (*N,N*-diallyl, 2-chloroacetamide). For example, corn leaves contain an enzyme which detoxifies atrazine by conjugating it with GSH. The higher the activity of this enzyme, the greater is the resistance of the plant to the herbicide.[31,32] Some herbicide "antidotes" appear to act by increasing intracellular GSH concentrations in plants.[33] Glutathione also appears to be a major form by which "organic" sulfur is transported from the leaves to the roots of plants.[34,35] Plants take up the sulfate ion from the soil through the roots. It is then translocated through the stem and into the leaves, where chloroplasts reduce it into cysteine.[9] Some of the cysteine is then converted into glutathione for export to other tissues of the plant.

C. Formation of Hydrogen Peroxide and Methods for Its Removal

Most oxygen taken up by aerobic cells is reduced to water by the addition of four electrons to each molecule

$$O_2 + 4e^- + 4H^+ \longrightarrow 2H_2O \qquad (5)$$

This reaction is catalyzed by the cytochrome oxidase complex of the inner mitochondrial membrane. Some enzymes, however, are oxidases which transfer two electrons onto each O_2 molecule that they use, making hydrogen peroxide:

$$O_2 + 2H^+ + 2e^- \longrightarrow H_2O_2 \qquad (6)$$

Important enzymes of this type in some plant tissues include urate oxidase, amino acid oxidases, and especially glycollate oxidase, an enzyme of the photorespiratory pathway (Figure 6). Some of the electrons passing down the electron transport chains of mitochondria from both green[36] and nongreen[37,38] plant tissues appear to leak away to oxygen en route, with the eventual production of H_2O_2.[36-38] Isolated illuminated chloroplasts produce H_2O_2, and CO_2 fixation by such preparations is often stimulated by addition of catalase,[39-40] indicating production of H_2O_2 during photosynthesis to an extent that can inhibit CO_2 fixation. Indeed, if H_2O_2 is added to the reaction medium, CO_2 fixation by isolated chloroplasts is 50% inhibited by concentrations as low as 10^{-5} *M*.[41,42] The site of inhibition seems to be the fructose and sedoheptulose diphosphatase enzymes[43,44] of the Calvin cycle which are oxidized by H_2O_2 to forms that are much less active under physiological conditions.[15] Oxygen uptake and H_2O_2 production have also been observed on illumination of intact cells of certain algae.[45,46]

Some authors have been unable to detect H_2O_2 production by intact chloroplasts in vitro, however.[47-49] In view of the powerful inhibitory effects of H_2O_2 on the Calvin cycle, one would expect the chloroplast to possess some protective mechanisms against it. If these mechanisms were of variable efficiency in the chloroplast preparations studied by different workers, this could explain the different results obtained. Let us now examine the possibilities.

Almost all plant tissues contain high activities of the heme-containing enzyme catalase. The purified *Lens culinaris* and spinach-leaf enzymes are tetramers of identical subunits.[50,51] Catalase breaks down high concentrations of H_2O_2 very rapidly, but it is much less effective at removing H_2O_2 present in low concentrations because of its low affinity (high K_m) for this substrate.[8,9,163] Most, if not all, of the catalase in leaf cells is located in peroxisomes, where it serves to remove H_2O_2 generated during the rapid oxidation of glycollate that occurs in photorespiration (Figure 6). Peroxisomes are present in most other plant tissues as well. Although much of the catalase activity present in homogenates of plant tissues is not associated with any of the particulate fractions obtained by centrifugation, this is thought to be due to rupture of fragile peroxisomes during the rather rigorous homogenization procedures needed to break the tough plant cell walls.[52] Variable amounts of catalase activity can be detected in isolated chloroplast preparations, but this is due to cytoplasmic contamination and little, if any, catalase is present in the stroma.[53] Small amounts of catalase activity have been reported to be associated with isolated photosystem I particles,[54] but this can be attributed to the fact that several of the other heme proteins present in the preparation can nonspecifically degrade H_2O_2 at low rates.[55] Hence catalase cannot be involved in the removal of H_2O_2 generated by illuminated chloroplasts or by mitochondria in vivo.

Plant tissues can also dispose of H_2O_2 by using peroxidase enzymes, which catalyze H_2O_2-dependent oxidation of substrates (SH_2) according to the general equation

$$SH_2 + H_2O_2 \longrightarrow S + 2H_2O \qquad (7)$$

Animal cells contain the enzyme glutathione peroxidase,[19,29] a selenium-containing protein which is specific for GSH as substrate

$$2GSH + H_2O_2 \longrightarrow GSSG + 2H_2O \qquad (8)$$

There is no evidence for the existence of selenium-containing glutathione peroxidases in plants,[215] but homogenates of leaf tissue contain an unstable enzyme activity which brings about removal of H_2O_2 in the presences of GSH.[56] A similar activity has been detected in isolated spinach chloroplast fractions,[57] although it might have been due to contamination of the preparation by other subcellular fractions.[9] Elucidation of the nature of this enzyme must await further studies.

However, plants do contain a wide range of peroxidases with a very broad specificity for substrate (SH_2). Such enzymes are usually detected in tissue extracts by using artificial substrates such as guaiacol or *o*-dianisidine. Perhaps the most studied peroxidase enzyme in biochemistry comes from a plant tissue, viz., horseradish (*Armoracia lapathifolia*) peroxidase. Oxidations by this and probably most other plant peroxidases can be represented by the following series of equations (for reviews see References 58 and 59).

$$\text{Peroxidase} + H_2O_2 \longrightarrow \text{compound I} \quad (9)$$

$$\text{Compound I} + SH_2 \longrightarrow SH\cdot + \text{compound II} \quad (10)$$

$$\text{Compound II} + SH_2 \longrightarrow SH\cdot + \text{peroxidase} \quad (11)$$

$$SH\cdot + SH\cdot \longrightarrow S + SH_2 \quad (12)$$

The iron in the heme ring of "resting" peroxidase is in the oxidation state Fe (III). H_2O_2 removes two electrons from the enzyme to give compound I, which has a nominal oxidation state of Fe (V). In fact, the iron is probably in the iron (IV) state and the extra oxidizing capacity is located elsewhere on the molecule. The two electrons are replaced in two one-electron steps (Equations 10 and 11), in each of which a molecule of substrate (SH_2) is oxidized to a radical SH. Sometimes the radical species can themselves reduce O_2 directly, leading to production of H_2O_2 in the reaction mixture, and so only catalytic quantities of H_2O_2 need be present to start the reaction.[58,59] For example, horseradish peroxidase will oxidize the reduced form of nicotinamide adenine dinucleotide (NADH) into the oxidized form of nicotinamide adenine dinucleotide (NAD^+)[59,60] and 2-nitropropane into acetone and nitrite[62] without its being necessary to add H_2O_2, since sufficient amounts of H_2O_2 are present in solutions of these substrates to start Reaction 9, and the substrate-derived radicals produced by Reactions 10 and 11 can interact directly with O_2 to generate more H_2O_2. In most cases, however, stoichiometric quantities of H_2O_2 are required to achieve substrate oxidation.

Peroxidases are involved in the degradation of the plant hormone indoleacetic acid[58,164] and thus in the regulation of plant growth. They also play a role in lignification. Lignification of plant cell walls involves the oxidative condensation of various phenols derived originally from cinnamic acid, and the role of peroxidase is to convert these phenols into the phenoxy radicals which actually polymerize.[63,64] How the H_2O_2 required by the peroxidase is generated by the cell walls is not clear. One possibility is the oxidation by peroxidase of NADH generated by a malate dehydrogenase enzyme bound to the cell wall,[65,66] which produces surplus H_2O_2 as explained above. Another suggestion is that glycollate oxidase (Figure 6) is present in woody tissue, but the glyoxylate it produces is recycled back to glycollate by a glyoxylate reductase enzyme for further oxidation and H_2O_2 generation.[67,68]

However, the identity of the natural substrates of most plant peroxidases is unknown so that it is difficult to assess their contribution to removal of H_2O_2 in vivo. Chemiluminescence (light emission) by root and stem tissues from a wide range of plants has been detected by use of liquid scintillation counting and it has been suggested to be due to the interaction of H_2O_2 with peroxidases in vivo,[69] although other sources are possible.[70]

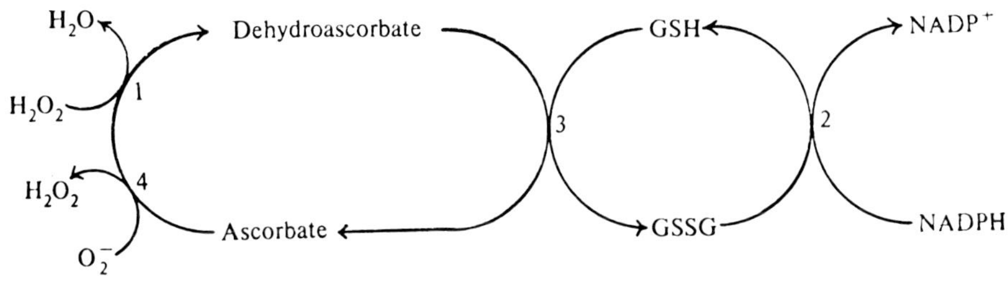

FIGURE 8. An ascorbate-glutathione cycle for removal of H_2O_2 in illuminated chloroplasts. Enzymes involved: (1) ascorbate peroxidase; (2) glutathione reductase; (3,4) nonenzymic reactions. Reaction 3 proceeds rapidly at pH 8, the pH of the stroma in illuminated chloroplasts. Leaf tissues contain a dehydroascorbate reductase enzyme, which catalyzes this reaction at acidic and neutral pH values, but it is not located within chloroplasts (see text).

Isolated chloroplasts contain little, if any, "nonspecific" peroxidase activity.[71] However, Groden and Beck[72] have detected an ascorbate peroxidase in spinach chloroplasts, which catalyzes the reaction

$$\text{Ascorbate} + H_2O_2 \longrightarrow \text{dehydroascorbate} + 2H_2O \qquad (13)$$

Since the chloroplast stroma contains ascorbate at concentrations up to 50 mM,[73-75] ascorbate would be a likely candidate for the role of H_2O_2 removal. At pH 8, the pH of the stroma in the illuminated chloroplast,[9] dehydroascorbate is rapidly reduced by GSH in a nonenzymic reaction[17,76] and the GSSG so produced can be reduced by chloroplast glutathione reductase. It is therefore possible that a cycle of reactions operates in chloroplasts to achieve removal of H_2O_2 at the expense of oxidizing NADPH from the electron transport chain. This cycle is shown in Figure 8. The green alga *Euglena*, which contains no catalase activity, also has the enzymes necessary to operate an ascorbate/glutathione cycle.[77,78] Leaf tissues have been reported to possess a soluble ascorbate peroxidase in addition to the chloroplast enzyme,[79] and the cytosol also contains an enzyme (dehydroascorbate reductase) which catalyzes reduction of dehydroascorbate by GSH at acidic and neutral pH values.[76,158,210-212] Such an enzyme is not necessary in illuminated chloroplasts, since this reaction can occur nonenzymically at alkaline pH values.

Both the ascorbate content of chloroplasts and their ascorbate peroxidase activity vary over a wide range, depending on the plant studied and the time of year.[72,75] Variations in glutathione concentration might also occur.[80] In chloroplasts with a low ascorbate and/or glutathione content, the cycle shown in Figure 8 would be impaired, H_2O_2 could accumulate during photosynthesis, and it would be necessary to add catalase to the reaction medium to obtain maximum rates of CO_2 fixation. Indeed, Egneus et al.[39] found that added ascorbate was able to replace catalase in stimulating CO_2 fixation by their chloroplast preparations. In preparations with high stromal ascorbate and GSH concentrations, H_2O_2 should not accmulate and the effects of catalase reported by Egneus et al.[39] and by others would not be expected to occur.

D. Lipid Peroxidation and the Formation of Singlet Oxygen

Lipids make up about 35% of the dry weight of chloroplasts. The major lipids present are glycolipids together with some sulfoquinovosyl diglyceride, often called "the plant sulfolipid" (Figure 9). The lipid compositions of envelope and thylakoid membranes are different.[9] Most of the fatty acids present in an esterified form in chloroplast lipids are C_{18} acids containing one or more double bonds. α-Linolenic acid ($C_{18:3}$) can sometimes comprise over 90% of the esterified fatty acids.[9] An unsaturated fatty

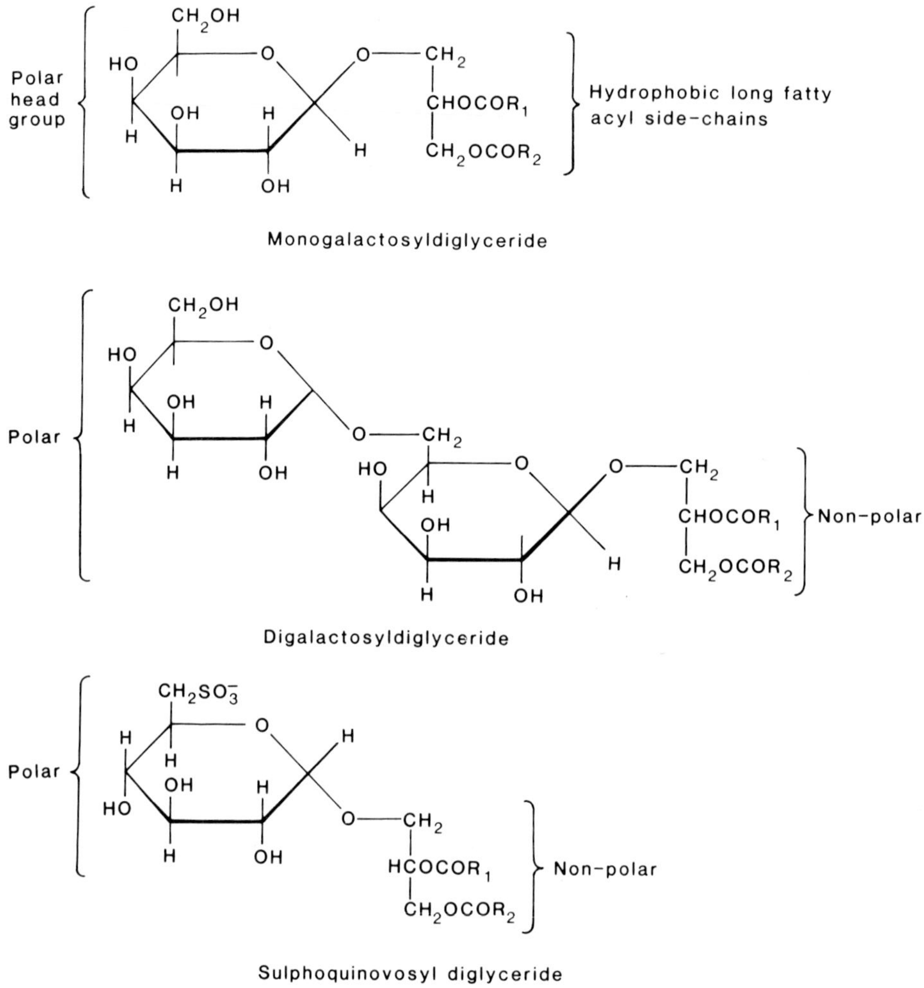

FIGURE 9. Structure of the chloroplast galactolipids and of sulfoquinovosyl diglyceride.

acid unique to photosynthetic tissues is *trans*-3-hexadecenoic acid ($C_{16:1}$), which occurs acylated to phosphatidylglycerol (Figure 10).

Polyunsaturated fatty acid side chains are prone to undergo a process known as lipid peroxidation, which involves the reaction of oxygen with conjugated dienes to produce lipid peroxides (Figure 11). Lipid peroxides decompose to give aldehydes (e.g., malondialdehyde) and other products, including volatile hydrocarbons such as ethane and pentane.[81,82] Their decomposition is greatly accelerated in the presence of transition metal ions or heme compounds. The peroxides and some of their degradation products cause extensive damage to enzymes and to membranes, producing a decrease in electrical resistance and membrane fluidity and eventual loss of membrane integrity.[83-85] Disruption of, for example, the tonoplast membrane of a plant cell by lipid peroxidation might release stored organic acids and hydrolytic enzymes from the central vacuole into the cytoplasm, so potentiating the damage. The loss of viability of soybean seeds that occurs during storage in warm, damp conditions has been attributed to accumulation of lipid peroxides,[86] and many of the damaging effects of elevated O_2 concentrations on plant tissues seem to be due to increased leakiness of membranes.[4] In contrast little, if any, lipid peroxide can be detected in extracts from plant tissues kept under

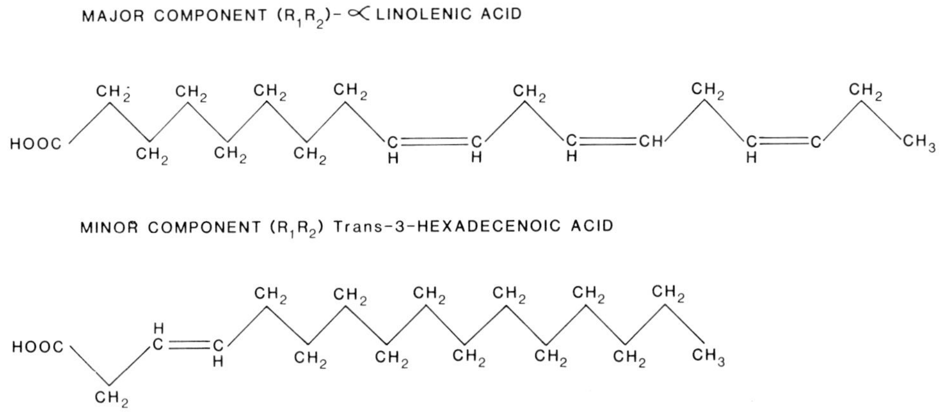

FIGURE 10. Fatty acids found in chloroplast lipids. α-Linolenic acid is a major component of lipids in chloroplasts of all higher plants and in some species comprises over 90% of all esterified fatty acids present. Note the high degree of unsaturation and the *cis*configuration of the double bonds.

normal conditions, which suggests the occurrence of protective mechanisms that become overwhelmed at high O_2 concentrations.

The rate of peroxidation of pure lipids or fatty acids is very slow indeed, but it is greatly accelerated by the presence of transition metal ions (especially iron or copper),[142,166] heme compounds or illuminated chlorophyll.[142] Once initiated, peroxidation is autocatalytic (Figure 11). Lipid peroxidation may be controlled by the presence of "scavengers" which trap the peroxide radicals and so interrupt the chain reaction (Figure 11). Leaf tissues contain large amounts of the scavenger α-tocopherol (Figure 12), known to animal nutritionists as vitamin E. Its lipophilic nature enables it to deal effectively with peroxides formed inside membranes. Most of the α-tocopherol of leaf tissues is located in chloroplasts,[165] being particularly concentrated in photosystem I.[87] The radical form of this compound (α-T· in Figure 11) can be reduced back to α-tocopherol by ascorbic acid,[88] which is present in the stroma at high concentrations and could therefore interact with tocopheryl radicals at the membrane surface, so as to regenerate tocopherol for further use in preventing lipid peroxidation.

Prolonged illumination of chloroplast thylakoids in vitro causes marked lipid peroxidation. Indeed, when polyunsaturated fatty acids are mixed with chlorophyll and illuminated, they are rapidly peroxidized.[89-91] This happens because the higher electornic excitation states formed on illumination of the chlorophyll molecule (Figure 3) are capable of transferring their excess energy onto the oxygen molecule, raising it from its ground state to a more-reactive excited state known as singlet oxygen $^1\Delta_g$ or $^1O_2$. The higher excited state of oxygen ($^1\Sigma g^+$) may also be formed, but it usually decays to the lower state $^1\Delta_g$ before it has time to react with anything.

Singlet oxygen $^1\Delta_g$ has a very short lifetime in aqueous solution, but it survives much longer in hydrophobic environments, such as the interior of membranes. Unlike ground-state oxygen, singlet oxygen can react directly with polyunsaturated fatty acid side chains to form lipid peroxides (Figure 11). Since the trapping of light by photosystems I and II causes formation of the excited states of the reaction center chlorophyll molecules and chloroplasts have a high internal O_2 concentration, production of singlet oxygen might be expected whenever there is "excess" excitation, i.e., when operation of the electron transport chain is restricted so that the electron acceptors of the two photosystems become saturated with electrons. This accounts for the increase in the rate of lipid peroxidation observed in isolated illuminated chloroplasts or algal cells when electron transport is blocked by the herbicides diuron (DCMU) or monuron

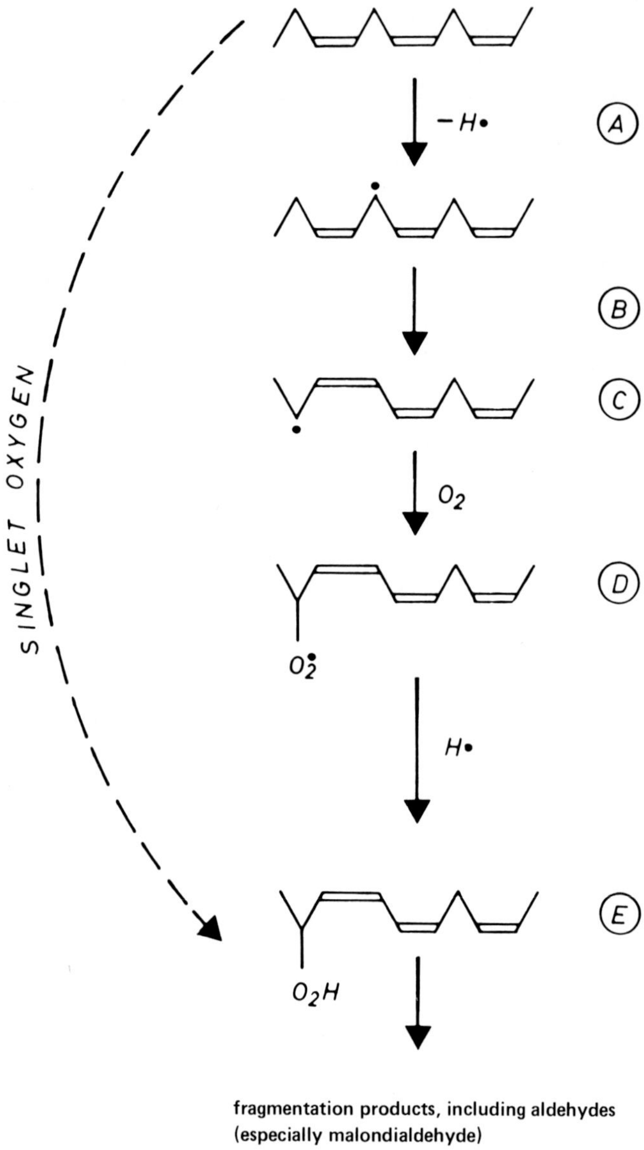

FIGURE 11. Mechanism of peroxidation of polyunsaturated fatty acids. The chain reaction of lipid peroxidation can be effectively inhibited by abstraction of ·H from another donor such as α-tocopherol (α-TH) and so such compounds inhibit lipid peroxidation.

$$\text{Lipid} - O_2^{\cdot} + \alpha\text{TH} \longrightarrow \text{Lipid} - \text{OOH} + \alpha - \text{T}\cdot$$

Singlet O_2 reacts directly with polyunsaturated fatty acids to give hydroperoxides. It has been suggested that breakdown of lipid peroxides also releases singlet oxygen. (A) Hydrogen abstraction, by a previously formed peroxide radical or by a species such as ·OH generated from $O_2^{\cdot-}$; (B) rearrangement; (C) conjugated diene (can be detected by absorbance at 233 nm); (D) peroxide radical: abstracts H· from another chain, thus causing a chain reaction; (E) hydroperoxide.

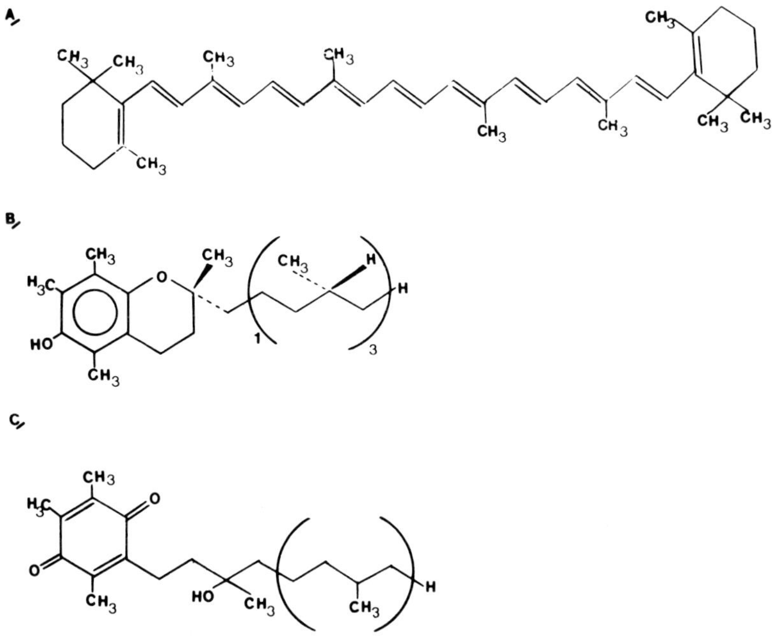

FIGURE 12. Structure of compounds involved in controlling lipid peroxidation in chloroplasts. (A) β-carotene. A scavenger of singlet O_2 and a quencher of excess chlorophyll excitation energy. (B) α-tocopherol. A scavenger of singlet O_2 and a "free radical trap" for lipid peroxides. About one molecule is present in spinach chloroplasts for every ten molecules of chlorophyll *a*. (C) α-tocopherylquinone. About 1 molecule is present in spinach chloroplasts for every 40 molecules of chlorophyll a.

(CMU) and the decrease in peroxidation induced by adding electron acceptors such as ferricyanide.[89,90,92,93] Singlet oxygen can also damage the chlorophyll molecule itself, causing loss of the characteristic green color ("bleaching"). Indeed, the chlorophyll bleaching observed in flax cotyledons treated with CMU has been shown to require both light and O_2.[93]

Since chloroplasts are likely to form singlet O_2 in vivo, one would expect them to be protected against it. As well as being an inhibitor of lipid peroxidation, α-tocopherol is a powerful scavenger of singlet O_2. The products of reaction of tocopherol with singlet O_2 include tocopherylquinone,[94] which may account for the presence of this compound in chloroplasts in significant amounts (Figure 12). Consistent with this suggestion, illumination of isolated spinach chloroplasts under conditions favoring lipid peroxidation has been shown to promote conversion of tocopherol to tocopherylquinone in the thylakoids.[95] It must not be assumed, however, that α-tocopherol in plant tissues serves only as a protection against membrane damage. Other roles are perhaps suggested by the fact that a tocopherol oxidase enzyme, whose activity is under hormonal control, has been detected in many plant tissues,[96,97] although this enzyme has so far resisted purification and its subcellular location is unknown.

Carotenoids, which are important constituents of chloroplast membranes, react with singlet oxygen at a diffusion-controlled rate and can therefore help to protect chlorophyll and membranes against damage.[98] Carotenoids are quickly destroyed on illumination of thylakoids in vitro, presumably as they absorb any singlet oxygen formed.[99,100] Their protective function in vivo is well illustrated by mutant strains of maize which lack them: illumination of such plants under aerobic conditions causes rapid bleaching of chlorophylls and destruction of chloroplast membranes. Illumina-

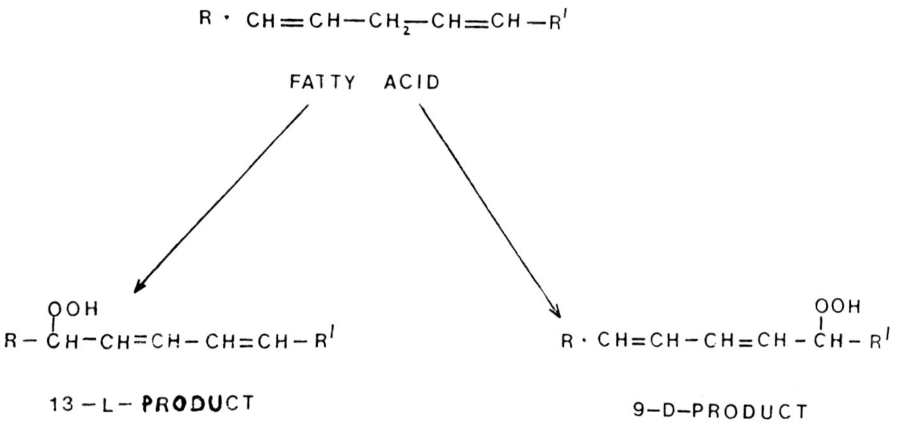

FIGURE 13. Reaction catalyzed by lipoxygenase. Lipoxygenases catalyze the peroxidation of fatty acids with a *cis, cis* 1,4-pentadiene structure. For example, linoleic acid may be converted to 13-L or to 9-D hydroperoxy-derivatives, depending on the enzyme. Some enzymes produce only one isomeric product whereas others give both.[113] The reactions are highly stereospecific.

tion under anaerobic conditions causes much less damage, since singlet oxygen cannot then be generated.[101] Similar destructive effects of illumination in the presence of O_2 are seen in normal plants in which carotenoid biosynthesis has been inhibited by herbicides.[102] The carotenoids present in chloroplasts are two main types, the carotenes (e.g., β-carotene, whose structure is shown in Figure 12) and xanthophylls, which are oxygen-containing derivatives of carotenes. Carotenoids are also able to quench directly those excited states of chlorophyll that lead to singlet oxygen formation. Hence they have a dual role: preventing formation of singlet oxygen in vivo and helping to remove any that does happen to be formed.[103,104] In addition to α-tocopherol and carotenoids, ascorbate[105] and glutathione are also rapidly oxidized by singlet oxygen and might help to remove any that is formed in the chloroplasts.

The deterioration of isolated, illuminated chloroplast thylakoids is in fact an extremely complicated process. In addition to lipid peroxidation, aldehyde formation, and carotenoid destruction, there is actual hydrolysis of glycolipids and phospholipids to release fatty acids. Both esterified and released fatty acids can undergo lipid peroxidation.[106,107] Galactolipid hydrolysis seems to be due to the action of galactolipase enzymes, which normally show little activity in isolated chloroplasts but seem to be "unmasked" during membrane deterioration.[108,109] The free fatty acids produced by lipase action themselves cause severe membrane damage[110] and inhibit photosynthesis.[111,112]

The ascorbate peroxidase enzyme of *Euglena*, which plays an important role in the removal of H_2O_2 as discussed in the previous section, is also capable of reducing artifical organic hydroperoxides.[78] It might therefore be capable of reducing lipid peroxides in vivo, and further experiments are required on this point.

Lipoxygenase (a useful role for lipid peroxidation?) — Many nongreen plant tissues, such as tubers, fruits, and seeds, contain the enzyme lipoxygenase, an iron-containing protein which catalyzes a direct reaction of polyunsaturated fatty acids with O_2 to give 13- and 9-hydroperoxides (Figure 13).[113,114] Low activities of this enzyme are sometimes present in leaf tissues.[115] Wounding of plant tissues such as cucumbers or potato tubers initiates a series of reactions that result in lipid peroxidation and subsequent degradation of the peroxides.[113] The enzymes involved are hydrolases, which release free fatty acids from membrane-bound lipids, lipoxygenase to produce fatty acid peroxides, and

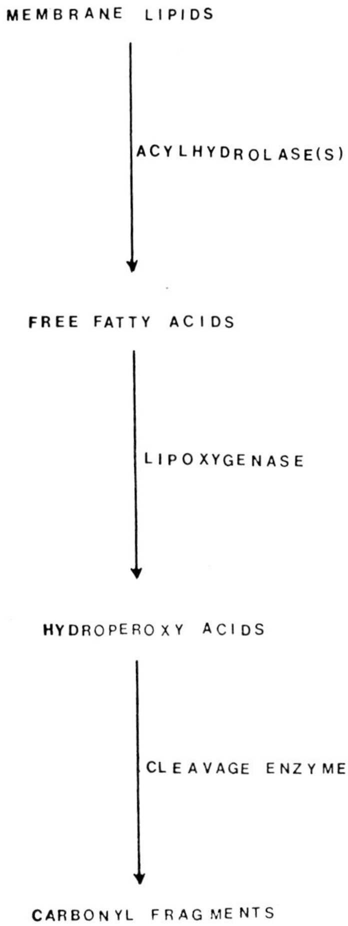

FIGURE 14. Pathway for the breakdown of membrane lipids induced on wounding plant tissues.

enzymes for the further metabolism and cleavage of these compounds (Figure 14). Some of the aldehydes so produced have characteristic smells, which are responsible for the aroma of certain plant tissues, such as sliced cucumbers and tea. The odor of crushed green leaves is caused by a process similar to that shown in Figure 14.[116,117] Chloroplast fractions from leaves of tea and Japanese silver (*Farfugium japonicum*) plants convert C_{18} unsaturated fatty acids into C_6-aldehydes, but the role of lipoxygenase in the reactions has not been established, nor is it rigorously proven that the activities are actually located within the chloroplast and not due to contamination with other organelles.[117-119] The formation of lipid peroxides and aldehydes when plant tissues are damaged may play an important role in killing fungi and bacteria attempting to enter the wound. Indeed, crushed leaves of Japanese silver are a traditional remedy for wound infections. It has also been speculated that some of the compounds formed may be insect attractants or repellants.[113,120] Damage to certain plant tissues can result in formation of "traumatic acid" (wound hormone), a compound which induces cell proliferation. Traumatic acid is formed by oxidative degradation of polyunsaturated fatty acids by processes related to those discussed above.[113] Perhaps we are looking here at an example of controlled lipid peroxidation put to useful purposes by plant tissues.

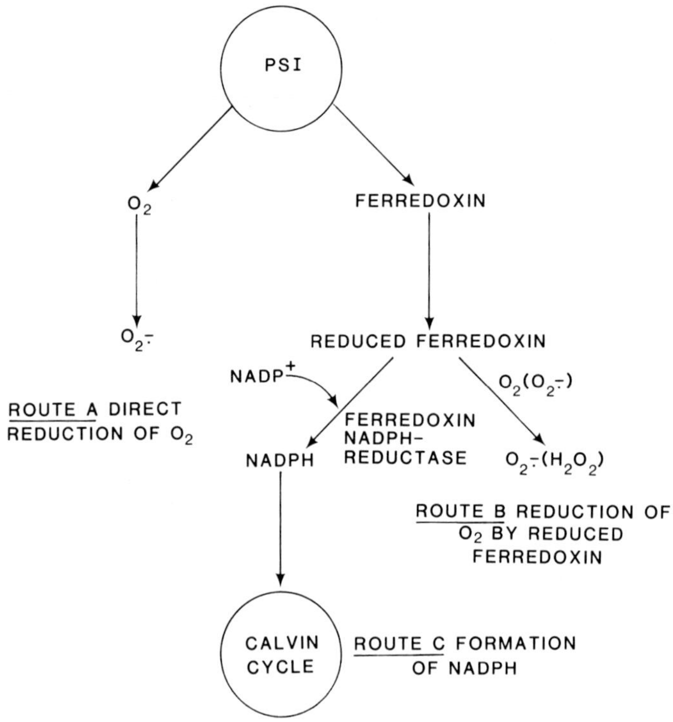

FIGURE 15. Possible pathways of electron flow from reduced photosystem I.

E. Formation of Oxygen Free Radicals

Some cellular systems catalyze oxidation reactions in which a single electron is transferred from the substrate onto each molecule of O_2 used, so producing an oxygen free radical known as superoxide, $O_2^{\cdot -}$ (Equation 14). The term "radical" can be applied to any species that has an unpaired electron.

$$O_2 + e^- \longrightarrow O_2^{\cdot -} \tag{14}$$

Examples of superoxide-producing enzymes present in some plant tissues are nitropropane dioxygenase,[121] galactose oxidase,[122] and xanthine oxidase.[123] Indeed, a mixture of xanthine and xanthine oxidase is often used as a source of $O_2^{\cdot -}$ for experiments in vitro.[123] "Nonspecific" plant peroxidases can generate $O_2^{\cdot -}$ during oxidation of certain substrates, e.g., NADH[58,61] and 2-nitropropane.[62]

The autoxidation of some reduced compounds also proceeds with single electron transfer to oxygen to give $O_2^{\cdot -}$. Such compounds include reduced flavins, pteridines, diphenols, and ferredoxin[124,214] (Equation 15).

$$\text{Ferredoxin}_{\text{reduced}} + O_2 \longrightarrow \text{ferredoxin}_{\text{oxidized}} + O_2^{\cdot -} \tag{15}$$

Allen[125] has proposed that reduced ferredoxin has a sufficiently low redox potential to further reduce $O_2^{\cdot -}$ to H_2O_2 (Equation 16), but this reaction has not yet been rigorously proved to occur.

$$\text{Ferredoxin}_{\text{reduced}} + O_2^{\cdot -} + 2H^+ \longrightarrow H_2O_2 + \text{ferredoxin}_{\text{oxidized}} \tag{16}$$

Isolated illuminated chloroplast thylakoids slowly take up O_2 in the absence of added electron acceptors. This was first observed by Mehler[126] and is hence often referred to as the "Mehler reaction". It appears to result from the reduction of O_2 to $O_2^{\cdot-}$ by electron acceptors associated with photosystem I.[127-130,206] Addition of ferredoxin increases the amount of $O_2^{\cdot-}$uptake, since it is reduced by photosystem I (PSI) much more quickly than is O_2 and the reduced ferredoxin can then undergo Reaction 15 and possibly Reaction 16. In vivo, however, reduced ferredoxin also passes electrons onto $NADP^+$ via ferredoxin-$NADP^+$ reductase (Figure 4). Thus electrons from photosystem I can pass through at least three routes, as shown in Figure 15, of which Route C is preferred.[130] If the supply of $NADP^+$ were limited, however, the rate of electron flow along Pathway C would be expected to be decreased and more $O_2^{\cdot-}$ should be made by Route B and, to a lesser extent, by Route A (Figure 15).[131,45,167] In some experiments, however, isolated intact chloroplasts have been shown to reduce O_2 even when light, rather than CO_2, was limiting the rate of photosynthesis.[39]

Elstner et al.[132,133] and others[134] have suggested that another compound which they have isolated from chloroplast fractions is required to mediate ferredoxin-dependent O_2 uptake. It has been called "oxygen reducing factor" and seems to be a phenolic compound, possibly a derivative of 4-hydroxycinnamic acid. Chloroplasts contain enzymes that hydroxylate monophenols to give diphenols, which can then be oxidized into quinones.[9] Both diphenols and quinones react rapidly with $O_2^{\cdot-}$ to produce semiquinones,[124] and so it is not surprising that effects on O_2 uptake can be demonstrated when such compounds are added to illuminated chloroplasts in vitro. Whether these effects have physiological significance remains to be determined. Pteridines have been detected in chloroplast preparations.[135,136] Since reduced pteridines autoxidize to yield $O_2^{\cdot-}$,[137] they could also be involved in O_2 uptake by chloroplasts, although direct evidence for this is lacking.

Mitochondria from nongreen plant tissues also produce $O_2^{\cdot-}$, probably because of direct reduction of O_2 by one of the components of the electron transport chain.[37,38] The endoplasmic reticulum of animal cells is an additional source of $O_2^{\cdot-}$, which originates from the cytochrome P450 system.[124] Since the microsomal fraction obtained from some plant tissues carries out hydroxylation reactions dependent on cytochrome P450, e.g., the hydroxylation of cinnamic acid,[138] $O_2^{\cdot-}$ is likely to be produced from this source in vivo in plant cells as well.

1. Properties of Superoxide

$O_2^{\cdot-}$ in aqueous solution can act either as a reducing agent, giving up its extra electron, or as a weak oxidizing agent, becoming reduced to H_2O_2. For example, it can reduce cytochrome *f* (Equation 17) and plastocyanin (Equation 18), components of the chloroplast electron transport chain (Figure 4), but it oxidizes ascorbic acid (Equation 19) and possibly ferredoxin (Equation 16 above).

$$\text{Cytochrome } f\,(Fe^{3+}) + O_2^{\cdot-} \longrightarrow \text{cytochrome } f\,(Fe^{2+}) + O_2 \qquad (17)$$

$$\text{Plastocyanin } (Cu^{2+}) + O_2^{\cdot-} \longrightarrow \text{plastocyanin } (Cu^{+}) + O_2 \qquad (18)$$

$$2H^+ + 2O_2^{\cdot-} + \text{ascorbate} \longrightarrow 2H_2O_2 + \text{dehydroascorbate} \qquad (19)$$

Superoxide oxdizes α-tocopherol and thiol compounds such as glutathione, but these reactions are slow.[124,139,168] It also appears to convert the metal ion Mn^{2+} into a more reactive species. For example, $O_2^{\cdot -}$ is insufficiently reactive to oxidize NADH but if Mn^{2+} is added to the system there is a rapid oxidation o the co-factor. Apparently the Mn^{2+} reacts with $O_2^{\cdot -}$ to form a complex (probably MnO_2^+) that oxidizes the NADH.[140] Perhaps this or similar reactions might be involved in the water-splitting reactions associated with photosystem II, in which manganese is intimately involved.[9]

By comparison with other oxygen radicals, however, $O_2^{\cdot -}$ is rather unreactive. For example, it cannot react directly with membrane lipids to cause peroxidation.[124,141] Indeed, most $O_2^{\cdot -}$ formed in biochemical systems probably reacts with itself as shown in Equation 20.

$$O_2^{\cdot -} + O_2^{\cdot -} + 2H^+ \longrightarrow H_2O_2 + O_2 \qquad (20)$$

This nonenzymic "dismutation" of $O_2^{\cdot -}$ produces O_2 and H_2O_2. The results of Asada et al.[127] indicate that the H_2O_2 produced by isolated illuminated chloroplasts arises from $O_2^{\cdot -}$ (generated as in Figure 15) by Reaction 20, and that the O_2 uptake demonstrated in the experiments of Egneus et al.[39] and others[45,131,167] was due to the reduction of O_2 to $O_2^{\cdot -}$ and subsequent dismutation to produce H_2O_2. H_2O_2 production by plant mitochondria probably also proceeds via the $O_2^{\cdot -}$ radical.[37,38]

There is considerable evidence to show that H_2O_2 and $O_2^{\cdot -}$ can react together in biochemical systems to form the hydroxyl radical, $\cdot OH$.[142,169] Pure H_2O_2 and $O_2^{\cdot -}$ will not react together at significant rates in vitro unless traces of iron salts are added, when the reactions shown in Equations 21 and 22 occur.[124,142] Of course, iron salts are present in all biochemical reagents and in extracts of plant tissues[143] so that these reactions can readily occur in vivo.

$$Fe^{3+} - \text{salt} + O_2^{\cdot -} \longrightarrow Fe^{2+} - \text{salt} + O_2 \qquad (21)$$

$$Fe^{2+} - \text{salt} + H_2O_2 \longrightarrow \cdot OH + OH^- + Fe^{3+} - \text{salt} \qquad (22)$$

$$\text{Net} \quad H_2O_2 + O_2^{\cdot -} \xrightarrow[\text{catalyst}]{\text{Fe-salt}} O_2 + \cdot OH + OH^- \qquad (23)$$

Reduced ferredoxin may be able to bring about generation of $\cdot OH$ from H_2O_2 by a related mechanism,[144] and it has been suggested that illuminated chlorophyll might be able to directly generate $\cdot OH$ radicals as well as singlet oxygen.[145]

Hydroxyl radicals are the most reactive species known to chemistry: they will attack and damage almost every molecule found in living cells.[146,141,142] For example, they can hydroxylate the purine and pyrimidine bases present in DNA and give rise to mutations, and they can abstract hydrogen radicals from membrane lipids and so trigger peroxidation (Figure 11).

There is some evidence to suggest that part of the oxygen formed during the breakdown of $O_2^{\cdot -}$ radicals is singlet oxygen $^1\Delta_g$. It is not at present clear whether singlet oxygen arises during Reaction 20 or during Reactions 21 and 22. Experimental results are very confused, largely because of the lack of specificity of many so-called "specific" scavengers of singlet oxygen.[142] The damaging effects of $^1\Delta_gO_2$ have already been discussed.

Although $O_2^{\cdot -}$ is scarcely pleasant for plant cells, its real danger therefore seems to lie in its ability to form $\cdot OH$ and possibly singlet oxygen. Protective mechanisms are therefore essential. Mechanisms of protection against singlet O_2 in chloroplasts have already been discussed.

2. *Superoxide Dismutase (SOD) in Plant Tissues*

Since the pioneering work of McCord and Fridovich,[123] it has been clearly established that all aerobic organisms contain enzymes, known as superoxide dismutases which greatly accelerate Reaction 20. $O_2^{\cdot -}$ is thus rapidly removed, which should prevent its conversion into $\cdot OH$ radicals. Further, the $O_2^{\cdot -}$ produced during the enzyme-catalyzed Reaction 20 is ground-state O_2, not singlet O_2.[124] The SOD content of tomato fruits has been reported to vary inversely with their susceptibility to the "sunscald" damage caused by exposure to strong sunlight,[147] which does suggest that SOD plays a role in protection against species generated in illuminated pigment systems.[216]

SOD was originally isolated from mammalian red blood cells and found to contain copper and zinc at its active site.[123] All SODs since discovered have also been found to be metalloproteins. The copper-zinc enzyme is found in all eukaryotic cells with the exception of some green algae, such as *Euglena*.[179,180] It is not generally found in prokaryotic cells,[142] which contain SODs with iron or manganese as the prosthetic group. The eukaryote *Euglena* contains both iron and manganese enzymes, an observation that may have considerable evolutionary significance.[177-180] Manganese enzymes have been purified and studied in detail from the red alga *Porphyridium cruentum*,[170] the luminous fungus *Pleurotus olearius*,[171] the blue-green alga *Plectonema boryanum*,[172] yeast,[189] and leaves of kidney bean.[173] Iron enzymes have been studied in several blue-green algae[172,174-176,203] as well as in *Euglena*.[177,178] There has been a recent report of the presence of an iron-containing SOD in leaves of the higher plant mustard, *Brassica campestris*.[148] Whatever the metal present, the enzyme catalyzes the same reaction.

Most of the SOD activity of green leaves is located in the chloroplasts, where it is present as a copper-zinc enzyme.[149-154] Some of it is bound to the thylakoids, from which it can be released by washing in hypotonic solutions, and the rest is apparently free in the stroma.[149,151] The purified SOD protein from spinach chloroplasts is a dimer of identical subunits, each of which has an active site containing one atom of copper and one of zinc.[150] Copper-zinc enzymes purified from pea and corn seedlings,[181] wheat germ,[182] bakers' yeast,[183] pea seeds,[184] the fungi *Neurospora crassa*[187] and *Fusarium oxysporum*,[204] and tomato fruits[147] had similar properties. In nongreen plant tissues, copper-zinc SOD may be largely present in the cell cytosol,[185,186] although rigorous subcellular fractionation experiments have not yet been reported. Although several reports of a manganese containing dismutase in chloroplast fractions have appeared, this activity appears to be due to cytoplasmic contamination.[149] However, disruption and subfractionation of photosystem II results in the release of a manganese-containing peptide that can react with $O_2^{\cdot -}$.[155] Whether this peptide and/or $O_2^{\cdot -}$ play any role in the water-splitting process has yet to be determined. Mitochondria isolated from spinach leaves,[149] maize seedlings,[188] and fungi *Dactylium dendroides*[205] and *N. crassa*[187] contain a manganese enzyme in the matrix. A small amount of copper-zinc SOD is probably also present in the space between the inner and outer mitochondrial membranes.[149,187]

Copper-zinc SODs are slowly but irreversibly inactivated by their product, H_2O_2.[190] Hence, for example, the chloroplast enzyme cannot remain active unless efficient methods for disposal of H_2O_2 are operative[172] (see Section I.C). Ascorbic acid which is present in the chloroplast stroma at high concentrations, reacts rapidly with $O_2^{\cdot -}$ [191] and can help to remove this species, as well as playing a role in the removal of H_2O_2 (Figure 8).

3. *Increased Superoxide Generation as a Phytotoxic Mechanism*

It is possible that the amounts of SOD and ascorbate present in chloroplasts are only sufficient to cope with the normal rate of $O_2^{\cdot -}$ generation in vivo, so that an increased rate would swamp them. This helps to account for the toxic effects of ele-

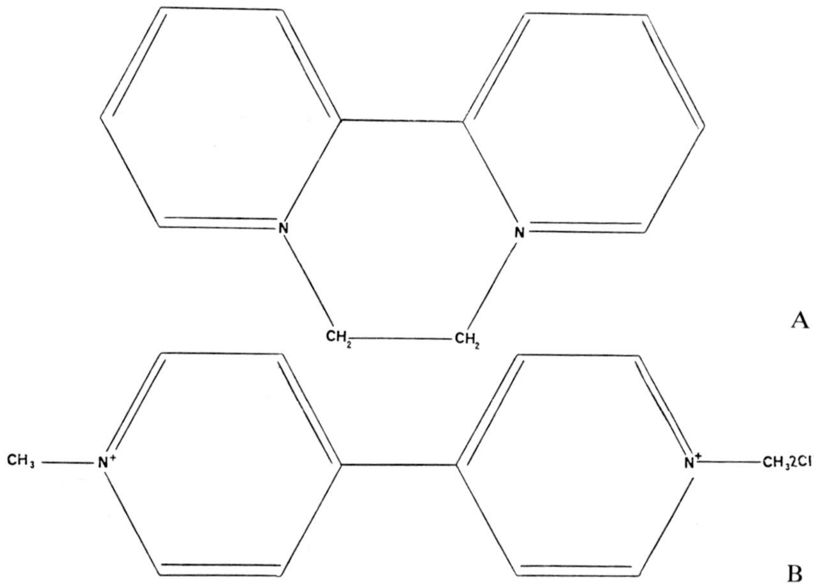

FIGURE 16. Structures of the bipyridyl herbicides (A) diquat and (B) paraquat.

vated O_2 concentrations, which presumably increase $O_2^{-\cdot}$ generation in chloroplasts because of increased competition of O_2 for electrons from photosystem I (Figure 15). Consistent with this, a mutant of the green alga *Chlorella* with abnormally high SOD activity is much more resistant to elevated O_2 concentrations than the wild-type strains.[213] Hence any compound which increased $O_2^{-\cdot}$ generation in vivo at normal O_2 concentrations would be expected to be toxic. The pathways for disposal of H_2O_2 in the chloroplast must also be able to cope with the extra H_2O_2 generated by any increased activity of SOD (Figure 8).

The damaging effects of the bipyridyl herbicides paraquat and diquat (Figure 16) on plant tissues may indeed be due to an increased rate of generation of H_2O_2 and $O_2^{-\cdot}$ in vivo. Their severe toxic effects on plants are only seen in the presence of both light and O_2, and increase in extent if the O_2 concentration around the plant is raised.[192] Bipyridyl herbicides are taken up rapidly by the roots and enter the leaves. The chloroplast envelope is freely permeable to paraquat and diquat, which can be reduced directly and rapidly by photosystem I in the light to give bipyridyl radical ions, $BP^{+\cdot}$ (Equation 24). This cannot, of course, occur in the dark.

$$BP^{2+} + H_2O \xrightarrow{\text{light}} \tfrac{1}{2}O_2 + BP^{+\cdot} \tag{24}$$

Such radicals rapidly autooxidize by the transfer of single electrons to O_2, making $O_2^{-\cdot}$.[193]

$$BP^{+\cdot} + O_2 \longrightarrow BP^{2+} + O_2^{-\cdot} \tag{25}$$

Addition of paraquat to chloroplasts therefore induces a rapid light-dependent oxygen uptake.[194] The enormously increased generation of $O_2^{-\cdot}$, and hence of $\cdot OH$, H_2O_2 and (possibly) singlet oxygen derived from it, inhibits CO_2 fixation. This is followed by lipid peroxidation and distruption of the chloroplast membranes, which is easily observed by electron microscopy of sections of paraquat-treated leaves[195,196] (Figure 17). In agreement with the above mechanism, addition of the related compound benzyl

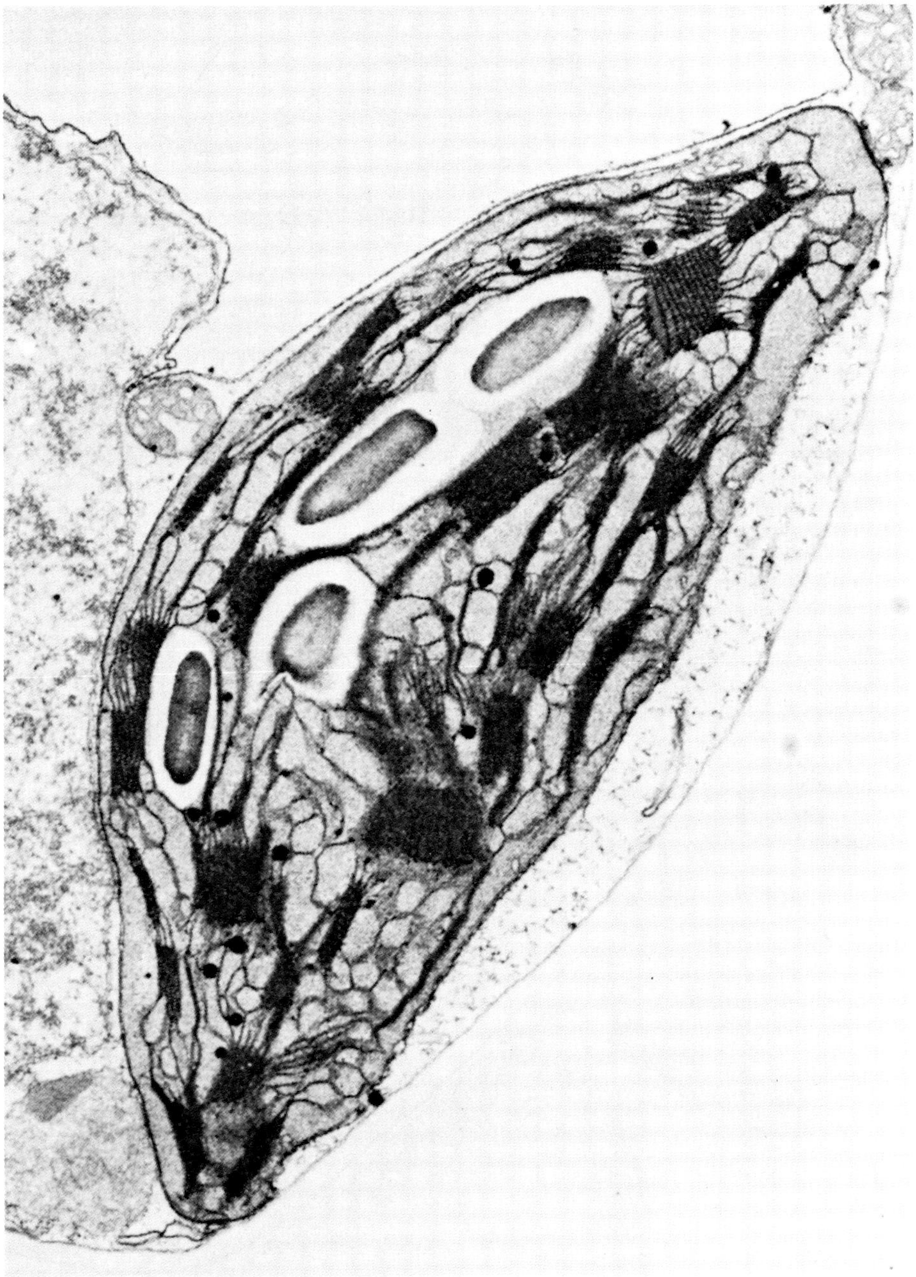

FIGURE 17. Effect of paraquat treatment of leaves on chloroplast structure. (A) The appearance of chloroplasts from leaves of ryegrass (*Lolium perenne* cultivar 'Kent Indigenous') under the electron microscope. (Magnification × 9000.) (B) 2 hr after paraquat treatment of the leaves the chloroplasts are swollen and the thylakoid membrane stacking has been extensively disrupted. By contrast the mitochondria are normal. (Magnification × 18,000.) (From Halliwell, B., *Chloroplast Metabolism. The Structure and Function of Chloroplasts in Green Leaf Cells,* 1st ed., Oxford University Press, Oxford, England, 1981. With permission.)

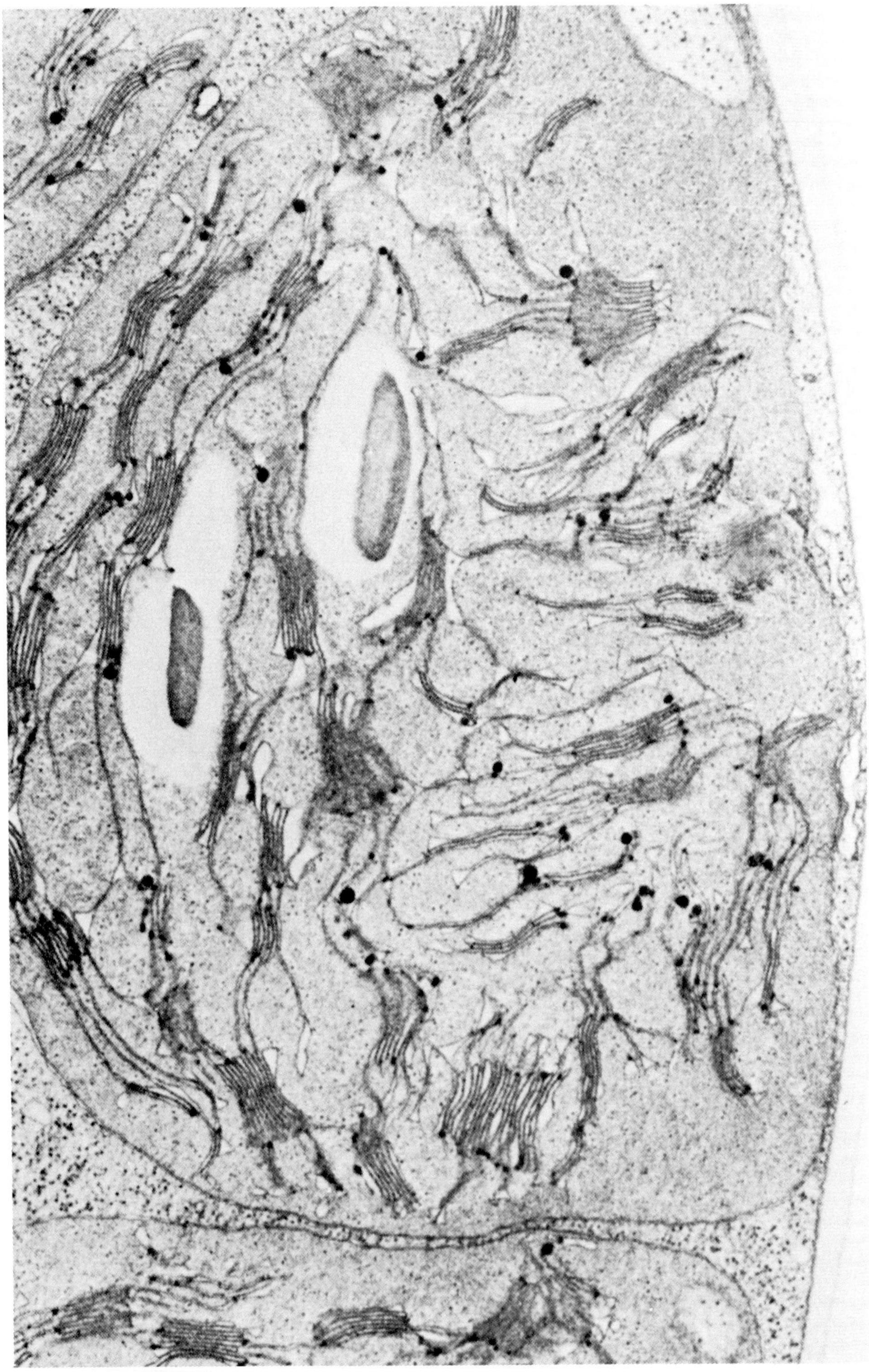

FIGURE 17B

viologen was found to increase the rate of peroxidation of isolated illuminated chloroplast thylakoids, and this increase was prevented by adding an excess of SOD.[197] A low-molecular-weight copper chelate that reacts with $O_2^{\cdot-}$ was found to decrease the toxicity of paraquat to flax cotyledons.[198] The leaves of some paraquat-tolerant strains of ryegrass are found to contain increased activities of SOD and catalase as compared with the leaves of sensitive plants.[199] The significance of the increase in catalase activity is not easy to evaluate, however, since this enzyme is not present within the chloroplasts.

II. SUMMARY: OVERALL SIGNIFICANCE OF OXYGEN EFFECTS IN PLANT TISSUES

Superoxide, H_2O_2, singlet oxygen, the hydroxyl radical ($\cdot OH$), and lipid peroxides are species whose formation is damaging to plant tissues and must therefore be carefully controlled. Chloroplasts are especially affected by them because of a high internal O_2 concentration, the presence of molecules which can reduce O_2 to $O_2^{\cdot-}$ (e.g., reduced ferredoxin), and the presence of pigments which can sensitize the formation of singlet O_2 and possibly of the hydroxyl radical. To allow their continued functioning, all plant tissues, but especially chloroplasts, have multiple protective mechanisms against these species. The main ones are SOD (removes $O_2^{\cdot-}$ and hence prevents formation of $\cdot OH$ and possibly, singlet O_2 from it), ascorbic acid (reacts rapidly with $O_2^{\cdot-}$, $\cdot OH$, and singlet O_2 and also helps remove H_2O_2 by the ascorbate peroxidase reaction), reduced glutathione (reacts rapidly with $\cdot OH$ and singlet oxygen, protects enzyme –SH groups, and helps to regenerate ascorbate from dehydroascorbate), α-tocopherol (inhibits the chain reaction of lipid peroxidation and scavenges singlet O_2 rapidly, and $O_2^{\cdot-}$ slowly), and carotenoids (decrease formation of singlet O_2 by absorbing excess excitation energy from chlorophyll by direct transfer, and also quench singlet O_2). Not until these systems are overloaded should toxic effects appear.

If spinach chloroplasts or isolated leaf cells are incubated in air from which CO_2 has been removed, their ability to fix CO_2 supplied to them subsequently is impaired.[200] It may be that in the absence of CO_2, no $NADP^+$ is available to accept electrons from photosystem I so that they are "shunted off" to O_2 (Figure 15). Any excess excitation of chlorophyll would cause more singlet oxygen to be formed. It has been suggested that the function of the photorespiratory pathway, which on first examination seems pointless in that it releases CO_2 from the plant which then has to be refixed with expenditure of more energy (Figure 6), is to prevent the damage that would result from excessive oxygen-radical production. By continually recycling CO_2, it should ensure that chloroplasts in vivo are never completely depleted of this molecule even when the leaves are exposed to CO_2-free air.

Formation of lipid peroxides and aldehydes on wounding of plant tissues may be an example of a situation where peroxidative reactions are put to a useful purpose. The ripening and senescence of fruits also seems to be a controlled oxidative process. As fruits ripen, there is a decrease in free –SH groups present and an increased formation of lipid peroxides and of H_2O_2. It seems that the progress of ripening is correlated with an increase in peroxide content of the fruit, which presumably leads to the breakdown of membranes and cell walls that is seen during ripening.[201,202,208] It is interesting to speculate that these changes could be brought about by the progressive withdrawal of the protective systems described above.

NOTE ADDED IN PROOF

There has been a recent report of the presence of iron-containing SOD in several higher plants in addition to *Brassica campestris*.[217]

ACKNOWLEDGMENTS

I am extremely grateful to the Rank Prize Funds, the Agricultural Research Council, and the Central Research Fund of the University of London for support of some of our research described in this review.

REFERENCES

1. **Poskuta, J., Mikulska, M., Faltynowicz, M., Bielak, B., and Worblewska, B.**, Effect of oxygen concentration on the development of photosynthetic capacity by etiolated bean seedlings upon illumination, *Z. Pflanzenphysiol.*, 7, 387, 1974.
2. **Marynick, M. and Addicott, F. T.**, Evidence for a dual role for oxygen in control of abscission, *Nature (London)*, 264, 668, 1976.
3. **Anderson, R. A. and Linney, T. L.**, Oxygen-dependent chemical tumorigenesis in a *Nicotiania* hybrid. Inhibition by ascorbic acid and by dinitrophenol, *Chem. Biol. Interact.*, 19, 317, 1977.
4. **Simon, E. W.**, Phospholipids and plant membrane permeability, *New Phytol.*, 73, 377, 1974.
5. **Steiger, H. M., Beck, E., and Beck, R.**, O_2 concentration in isolated chloroplasts during photosynthesis, *Plant Physiol.*, 60, 903, 1977.
6. **Gregory, R. P. F.**, *Biochemistry of Photosynthesis*, 2nd ed., John Wiley & Sons, London, 1977.
7. **Barber, J.**, Biophysics of photosynthesis, *Rep. Prog. Phys.*, 41, 1157, 1978.
8. **Halliwell, B.**, The chloroplast at work. A review of modern developments in our understanding of chloroplast metabolism, *Prog. Biophys. Mol. Biol.*, 33, 1, 1978.
9. **Halliwell, B.**, *Chloroplast Metabolism. The Structure and Function of Chloroplasts in Green Leaf Cells*, 1st ed., Oxford University Press, Oxford, England, 1981.
10. **Haugaard, N.**, Cellular mechanisms of oxygen toxicity, *Physiol. Rev.*, 48, 311, 1968.
11. **Morris, J. G.**, Nature of oxygen toxicity in anaerobic microorganisms, in *Strategies of Microbial Life in Extreme Environments*, Shilo, M., Ed., Verlag Chemie, Weinheim, 1979, chap. 10.
12. **Gomez-Moreno, C. and Ke, B.**, Oxidative inactivation of the molybdenum-iron-protein component of nitrogenase from *Clostridium pasteurianum*, *Mol. Cell. Biochem.*, 26, 111, 1979.
13. **Dilek Tozum, S. R. and Gallon, J.**, Effects of methyl viologen on *Gloeocapsa* sp. LB795 and their relationship to the inhibition of acetylene reduction (nitrogen fixation) by oxygen, *J. Gen. Microbiol.*, 111, 313, 1979.
14. **Ellyard, P. W. and Gibbs, M.**, Inhibition of photosynthesis by oxygen in isolated spinach chloroplasts, *Plant Physiol.*, 44, 1115, 1969.
15. **Charles, S. A. and Halliwell, B.**, Effect of hydrogen peroxide on spinach chloroplast fructose bisphosphatase, *Biochem. J.*, 189, 373, 1980.
16. **Rathnam, C. K. M. and Chollet, R.**, Photosynthetic carbon metabolism in C_4 plants and C_3-C_4 intermediate species, in *Progress in Phytochemistry*, Vol. 6, Harborne, J. B. and Swain, T., Eds., Pergamon Press, Oxford, chap. 1.
17. **Foyer, C. H. and Halliwell, B.**, Presence of glutathione and glutathione reductase in chloroplasts. A proposed role in ascorbic acid metabolism, *Planta (Berlin)*, 133, 21, 1976.
18. **Schaedle, M. and Bassham, J. A.**, Chloroplast glutathione reductase, *Plant Physiol.*, 59, 1011, 1977.
19. **Jocelyn, P. C.**, *Biochemistry of the Thiol Group*, 1st ed., Academic Press, London, 1972.
20. **Foyer, C. H.**, Function and Relationship of Ascorbic Acid and Glutathione in Spinach Leaves, Ph.D. thesis, University of London, King's College, England, 1977.
21. **Halliwell, B. and Foyer, C. H.**, Properties and physiological function of a glutathione reductase purified from spinach leaves by affinity chromatography, *Planta (Berlin)*, 139, 9, 1978.
22. **Mapson, L. W. and Isherwood, F.**, Glutathione reductase from germinated peas, *Biochem. J.*, 86, 173, 1963.
23. **Lendzian, K. and Bassham, J. A.**, Regulation of glucose-6-phosphate dehydrogenase in spinach chloroplasts by ribulose 1,5-diphosphate and NADPH/NADP$^+$ ratios, *Biochim. Biophys. Acta*, 396, 260, 1975.
24. **Hall, D. O.**, Coupling of photophosphorylation to electron transport in isolated chloroplasts, in *The Intact Chloroplast, Topics in Photosynthesis*, Vol. 1, Barber, J., Ed., Elsevier, Amsterdam, 1976, chap. 4.
25. **Mapson, L. W.**, Metabolism of ascorbic acid in plants: function, *Annu. Rev. Plant Physiol.*, 9, 119, 1958.

26. **Ernst, V., Levin, D. H., and London, I. M.**, Inhibition of protein synthesis initiation by oxidised glutathione. Activation of a protein kinase that phosphorylates the alpha-subunit of eukaryotic initiation factor 2, *Proc. Natl. Acad. Sci. U.S.A.*, 75, 4110, 1978.
27. **Isaccs, J. and Binkley, F.**, Glutathione-dependent control of protein-disulphide sulphydryl content by subcellular fractions of hepatic tissue, *Biochim. Biophys. Acta*, 497, 192, 1977.
28. **Kosower, N. S. and Kosower, E. M.**, The glutathione status of cells, *Int. Rev. Cytol.*, 54, 109, 1978.
29. **Chance, B., Sies, H., and Boveris, A.**, Hydroperoxide metabolism in mammalian organs, *Physiol. Rev.*, 59, 527, 1979.
30. **Diesperger, H. and Sandermann, H.**, Soluble and microsomal glutathione-S-transferase activities in pea seedlings *(Pisum sativum L.)*, *Planta (Berlin)*, 146, 643, 1979.
31. **Chasseaud, F.**, Nature and distribution of enzymes catalysing the conjugation of glutathione with foreign compounds, *Drug Metab. Rev.*, 2, 185, 1973.
32. **Shimabukuro, R. H., Lamoureux, G. L., and Frear, D. S.**, Glutathione conjugation. A mechanism for herbicide detoxication and selectivity in plants, in *Chemistry and Action of Herbicide Antidotes*, Pallos, F. M. and Casida, J. G., Eds., Academic Press, New York, 1978, 133.
33. **Lay, M. M. and Casida, J. E.**, Involvement of glutathione and glutathione-S-transferases in the action of dichloroacetamide antidotes for thiocarbamate herbicides, in *Chemistry and Action of Herbicide Antidotes*, Pallos, F. M. and Casida, J. G., Eds., Academic Press, New York, 1978, 151.
34. **Garsed, S. G. and Reid, D. J.**, Sulphur dioxide metabolism in soybean. Biochemical distribution of $^{35}SO_2$ products, *N. Phytol.*, 99, 583, 1977.
35. **Rennenberg, H., Schmitz, K., and Bergmann, L.**, Long-distance transport of sulphur in *Nicotiana tabacum*, *Planta (Berlin)*, 147, 57, 1979.
36. **Halliwell, B.**, Oxidation of formate by peroxisomes and mitochondria from spinach leaves, *Biochem. J.*, 138, 77, 1974.
37. **Huq, S. and Palmer, J. M.**, Superoxide and H_2O_2 production in cyanide-resistant *Arum maculatum* mitochondria, *Plant Sci. Lett.*, 11, 351, 1978.
38. **Rich, P. R. and Bonner, W. D.**, Sites of superoxide anion generation in higher plant mitochondria, *Arch. Biochem. Biophys.*, 188, 206, 1978.
39. **Egneus, H., Heber, U., Matthiesen, U., and Kirk, M.**, Reduction of O_2 by the electron transport chain of chloroplasts during assimilation of CO_2, *Biochim. Biophys. Acta*, 408, 252, 1975.
40. **Hind, G., Mills, J. D., and Slovacek, R. E.**, Cyclic electron transport in photosynthesis, in *Photosynthesis '77*, Hall, D. O., Coombs, J., and Goodwin, T. W., Eds., Biochemical Society, London, 1978, 591.
41. **Kaiser, W.**, Effect of H_2O_2 on CO_2 fixation of isolated intact chloroplasts, *Biochim. Biophys. Acta*, 440, 476, 1976.
42. **Robinson, J. M., Smith, M. G., and Gibbs, M.**, Influence of H_2O_2 upon CO_2 photoassimilation in the spinach chloroplast. H_2O_2 generated by broken chloroplasts in an intact chloroplast preparation is a causal agent of the Warburg effect, *Plant Physiol.*, 65, 755, 1980.
43. **Heldt, H. W., Chon, C. J., Lilley, R. McC., and Portis, A.**, Role of fructose and sedoheptulose bisphosphatase in the control of CO_2 fixation. Evidence from the effects of Mg^{2+} and H_2O_2, in *Photosynthesis '77*, Hall, D. O., Coombs, J., and Goodwin, T. W., Eds., Biochemical Society, London, 1978, 469.
44. **Kaiser, W.**, Reversible inhibition of the Calvin cycle and activation of oxidative pentose phosphate cycle in isolated intact chloroplasts by H_2O_2, *Planta (Berlin)*, 145, 377, 1979.
45. **Radmer, R. J. and Kok, B.**, Photoreduction of O_2 primes and replaces CO_2 assimilation, *Plant Physiol.*, 58, 336, 1976.
46. **Patterson, C. O. P. and Myers, J.**, Photosynthetic production of H_2O_2 by *Anacystis nidulans*, *Plant Physiol.*, 51, 104, 1972.
47. **Allen, J. F.**, Induction of a Mehler reaction in chloroplast preparations by methyl viologen and by ferredoxin. Effects on photosynthesis by intact chloroplasts, *Plant Sci. Lett.*, 12, 161, 1978.
48. **Allen, J. F.**, Induction of a Mehler reaction in chloroplast preparations by FMN. Effects on photosynthesis by intact chloroplasts, *Plant Sci. Lett.*, 12, 151, 1978.
49. **Allen, J. F. and Whatley, F. R.**, Effects of inhibitors of catalase on photosynthesis and on catalase activity in unwashed preparations of intact chloroplasts, *Plant Physiol.*, 61, 957, 1978.
50. **Gregory, R. P. F.**, An improved preparative method for spinach catalase and evaluation of some of its properties, *Biochim. Biophys. Acta*, 159, 429, 1968.
51. **Schiefer, S., Teifel, W., and Kindl, H.**, Plant microbody proteins. Purification and characterisation of catalase from leaves of *Lens culinaris*, *Hoppe-Seylers Z. Physiol. Chem.*, 357, 163, 1976.
52. **Tolbert, N. E.**, Microbodies — peroxisomes and glyoxysomes, *Annu. Rev. Plant Physiol.*, 22, 45, 1971.
53. **Allen, J. F.**, Effect of washing and osmotic shock on the catalase activity of intact chloroplast preparations, *FEBS Lett.*, 84, 221, 1977.

54. **Van Ginkel, G. and Brown, J. S.**, Endogenous catalase and superoxide dismutase activities in photosynthetic membranes, *FEBS Lett.*, 94, 284, 1978.
55. **Brown, R. H., Collins, N., and Merrett, M. J.**, Peroxidative activity in *Euglena gracilis*, *Plant Physiol.*, 55, 1123, 1975.
56. **Flohé, L. and Menzel, H.**, Influence of glutathione upon light-induced high-amplitude swelling and lipid peroxide formation of spinach chloroplasts, *Plant Cell Physiol.*, 12, 325, 1971.
57. **Wolosiuk, R. A. and Buchanan, B. B.**, Thioredoxin and glutathione regulate photosynthesis in chloroplasts, *Nature (London)*, 266, 565, 1977.
58. **Yamazaki, I. and Yokota, K.**, Oxidation states of peroxidase, *Mol. Cell. Biochem.*, 2, 39, 1973.
59. **Halliwell, B., Pasternack, R. F., and De Rycker, J.**, Interaction of the superoxide radical with peroxidase and with other haem compounds, in *Proc. 3rd Int. Symp. Oxidases and Related Redox Systems*, King, T. E., Mason, H. S., and Morrison, M., Eds., Pergamon Press, Oxford, 1982.
60. **Usumi, M., Matsushita, H., and Shimazu, T.**, Regulation of liver phosphorylase phosphatase by glutathione disulphide, *J. Biolog. Chem.*, 255, 1928, 1980.
61. **Halliwell, B.**, Lignin synthesis. Generation of H_2O_2 and superoxide by horseradish peroxidase and its stimulation by Mn^{2+} and phenols, *Planta (Berlin)*, 140, 81, 1978.
62. **De Rycker, J. and Halliwell, B.**, Oxidation of 2-nitropropane by horseradish peroxidase. Involvement of H_2O_2 and of superoxide in the reaction mechanism, *Biochem. J.*, 175, 601, 1978.
63. **Harkin, J. M. and Obst, J. R.**, Lignification in trees. Indication of exclusive peroxidase participation, *Science*, 180, 296, 1973.
64. **Stafford, H. A.**, Metabolism of aromatic compounds, *Annu. Rev. Plant Physiol.*, 25, 459, 1974.
65. **Gross, G. G., Janse, C., and Elstner, E. F.**, Involvement of malate, monophenols and the superoxide radical in H_2O_2 formation by isolated cell walls from horseradish, *Planta (Berlin)*, 136, 271, 1977.
66. **Mader, M., Ungemach, J., and Schloss, P.**, Role of peroxidase isoenzyme groups of *Nicotiana tabacum* in H_2O_2 formation, *Planta (Berlin)*, 147, 467, 1980.
67. **Sagisaka, S.**, Occurrence of peroxide in a perennial plant, *Populus gelrica*, *Plant Physiol.*, 57, 308, 1976.
68. **Sagisaka, S.**, Occurrence of NADP-linked glyoxylate reductase in nonphotosynthetic xylem tissue of perennials, *Plant Physiol.*, 65, 377, 1980.
69. **Abeles, F., Leather, G. R., and Forrence, L. E.**, Plant chemiluminescence, *Plant Physiol.*, 62, 696, 1978.
70. **Boveris, A., Sanchez, R. A., Varsavsky, A. I., and Cadenas, E.**, Spontaneous chemiluminescence of soybean seeds, *FEBS Lett.*, 113, 29, 1980.
71. **Parish, R. W.**, Intracellular location of phenoloxidases, peroxidases and phosphatases in leaves of spinach-beet, *Eur. J. Biochem.*, 31, 446, 1972.
72. **Groden, D. and Beck, E.**, H_2O_2 destruction by ascorbate-dependent systems from chloroplasts, *Biochim. Biophys. Acta*, 546, 426, 1979.
73. **Walker, D. A.**, Chloroplasts (and grana). Aqueous-including high carbon fixation ability, *Meth. Enzymol.*, 23A, 211, 1971.
74. **Gerhardt, B.**, Untersuchungen uber beziehungen zwischen ascorbinsaure und photosynthese, *Planta (Berlin)*, 61, 101, 1964.
75. **Groden, D. and Beck, E.**, Characterisation of a membrane-bound ascorbate-specific peroxidase from spinach chloroplasts, in *4th Int. Congr. Photosynthesis, Abstracts*, Coombs, J., Ed., United Kingdom Science Committee, London, 1977, 139.
76. **Foyer, C. H. and Halliwell, B.**, Purification and properties of dehydroascorbate reductase from spinach leaves, *Phytochemistry*, 61, 1347, 1977.
77. **Shigeoka, S., Nakano, Y., and Kitaoka, S.**, Metabolism of H_2O_2 in *Euglena gracilis Z.* by L-ascorbic acid peroxidase, *Biochem. J.*, 186, 377, 1980.
78. **Shigeoka, S., Nakano, Y., and Kitaoka, S.**, Purification and properties of L-ascorbic acid-specific peroxidase in *Euglena gracilis*, *Arch. Biochem. Biophys.*, 201, 121, 1980.
79. **Kelly, G. J. and Latzko, E.**, Soluble ascorbate peroxidase. Detection in plants and use in vitamin C estimation, *Naturwissenschaften*, 66, 5617, 1979.
80. **Esterbauer, H. and Grill, D.**, Seasonal variation of glutathione and glutathione reductase in needles of *Picea abies*, *Plant Physiol.*, 61, 119, 1978.
81. **Konze, J. R. and Elstner, E. F.**, Ethane and ethylene formation by mitochondria as an indication of aerobic lipid degradation in response to wounding of plant tissue, *Biochim. Biophys. Acta*, 528, 213, 1978.
82. **Pryor, W. A.**, Formation of free-radicals and the consequences of their reactions *in vivo*, *Photochem. Photobiol.*, 28, 787, 1978.
83. **Hicks, M. and Gebicki, J. M.**, A quantitative relationship between permeability and the degree of peroxidation in ufasome membranes, *Biochem. Biophys. Res. Commun.*, 80, 704, 1978.
84. **Pauls, K. P. and Thompson, J. E.**, *In vitro* stimulation of senescence-related membrane damage by ozone-induced lipid peroxidation, *Nature (London)*, 283, 504, 1980.

85. **Putvinsky, A. V., Sokolov, A. I., Roshcupkin, D. I., and Vladimirov, Y. A.,** Electric breakdown of bilayer phospholipid membrane under UV-induced lipid peroxidation, *FEBS Lett.*, 106, 53, 1979.
86. **Stewart, R. C. R. and Bewley, J. D.,** Lipid peroxidation associated with accelerated aging of soybean axes, *Plant Physiol.*, 65, 245, 1980.
87. **Baszynski, T.,** Effect of alpha-tocopherol on reconstitution of photosystem I in heptane-extracted spinach chloroplasts, *Biochim. Biophys. Acta*, 347, 31, 1974.
88. **Packer, T. E., Slater, T. F., and Willson, R. L.,** Direct observation of a free-radical interaction between vitamin E and Vitamin C, *Nature (London)*, 278, 737, 1979.
89. **Rawls, R. R. and Van Santen, P. J.,** Singlet oxygen. A possible source of the original hydroperoxides in fatty acids, *Ann. N.Y. Acad. Sci.*, 171, 135, 1970.
90. **Heath, R. L. and Packer, L.,** Photoperoxidation in isolated chloroplasts. Role of electron transfer, *Arch. Biochem. Biophys.*, 125, 850, 1968.
91. **Heath, R. L. and Packer, L.,** Photoperoxidation in isolated chloroplasts. Kinetics and stoichiometry of fatty acid peroxidation, *Arch. Biochem. Biophys.*, 125, 189, 1968.
92. **Elstner, E. F. and Osswald, W.,** Chlorophyll photobleaching and ethane production in DCMU- or paraquat-treated *Euglena gracilis* cells, *Z. Naturforsch.*, 35C, 129, 1980.
93. **Pallett, K. E. and Dodge, A. D.,** Role of light and O_2 in the action of the photosynthetic inhibitor herbicide monuron, *Z. Naturforsch.*, 34C, 1058, 1979.
94. **Clough, R. L., Yee, B. G., and Foote, C. S.,** Chemistry of singlet oxygen. The unstable primary product of tocopherol photooxidation, *J. Am. Chem. Soc.*, 101, 683, 1979.
95. **Yamauchi, R. and Matsushita, S.,** Light-induced lipid peroxidation in isolated chloroplasts and role of alpha-tocopherol, *Agric. Biol. Chem.*, 43, 2157, 1979.
96. **Barlow, S. M. and Gaunt, J. K.,** Metabolism of alpha-tocopherol by plants, *Phytochemistry*, 11, 2161, 1972.
97. **Gaunt, J. K., Matthews, G. M., and Plumpton, E. S.,** Control *in vitro* of tocopherol oxidase by light and by auxins, kinetin, gibberellic acid, abscisic acid and ethylene, *Biochem. Soc. Trans.*, 8, 186, 1980.
98. **Foote, C. S.,** Quenching of singlet oxygen, *Ann. N.Y. Acad. Sci.*, 171, 139, 1970.
99. **Takahama, U. and Nishimura, M.,** Formation of singlet molecular O_2 in illuminated chloroplasts. Effects on photoinactivation and lipid peroxidation, *Plant Cell Physiol.*, 16, 737, 1975.
100. **Ridley, S. M.,** Interaction of chloroplasts with inhibitors. Induction of chlorosis by diuron during prolonged illumination *in vitro*, *Plant Physiol.*, 59, 724, 1977.
101. **Anderson, I. C. and Robertson, D. S.,** Role of carotenoids in protecting chlorophyll from photodestruction, *Plant Physiol.*, 35, 531, 1960.
102. **Forsch, S., Jabben, M., Bergfeld, R., Kleinig, H., and Mohr, H.,** Inhibition of carotenoid biosynthesis by the herbicide SAN 9789 and its consequences for the action of phytochrome on plastogenesis, *Planta (Berlin)*, 145, 497, 1979.
103. **Foote, C. S. and Denny, R. W.,** Chemistry of singlet oxygen. Quenching by beta-carotene, *J. Am. Chem. Soc.*, 90, 6233, 1968.
104. **Krinsky, N. I.,** Carotenoid pigments. Multiple mechanisms for coping with the stress of photosensitised oxidations, in *Strategies of Microbial Life in Extreme Environments*, Shilo, M., Ed., Verlag Chemie, Weinheim, 1979, chap. 11.
105. **Bodannes, R. S. and Chan, P. C.,** Ascorbic acid as a scavenger of singlet O_2, *FEBS Lett.*, 105, 195, 1979.
106. **Harnischfeger, G.,** Photosensitised inhibitor formation in isolated aging chloroplasts, *Planta (Berlin)*, 104, 316, 1972.
107. **Hoshina, S. and Nishida, K.,** Photoswelling and light-inactivation of isolated chloroplasts. Functional and structural changes in isolated chloroplasts under the influence of lysolecithin, *Plant Cell Physiol.*, 16, 475, 1975.
108. **Anderson, M. M., McCarty, R. E., and Zimmer, E. A.,** Role of galactolipids in spinach chloroplast lamellar membranes. Partial purification of a bean leaf galactolipid lipase and its action on subchloroplast particles, *Plant Physiol.*, 53, 699, 1974.
109. **Wintermans, J. F. G. M., Helmsing, P. J., Polman, B. J. J., Van Gisbergen, J., and Collard, J.,** Galactolipid transformations and photochemical activities of spinach chloroplasts, *Biochim. Biophys. Acta*, 189, 95, 1969.
110. **Siegenthaler, P. A.,** Changes in pH dependence and sequential inhibition of photosynthetic activity in chloroplasts by unsaturated fatty acids, *Biochim. Biophys. Acta*, 305, 153, 1973.
111. **Golbeck, J. H., Martin, I. F., and Fowler, C. F.,** Mechanism of linolenic acid-induced inhibition of photosynthetic electron transport, *Plant Physiol.*, 65, 707, 1980.
112. **Mve Akamba, L. and Siegenthaler, P. A.,** Effect of linolenate on photosynthesis by intact spinach chloroplasts. Conditions for the inhibition of orthophosphate uptake, *FEBS Lett.*, 99, 6, 1979.

113. **Galliard, T.**, Lipolytic and lipoxygenase enzymes in plants and their action in wounded tissues, in *Biochemistry of Wounded Plant Tissues*, Kahl, G., Ed., Walter de Gruyter & Co., Berlin, 1978, 155.
114. **de Groot, J. J. M. C., Veldink, G. A., Vliegenthart, J. F. G., Boldingh, J., Wever, R., and Van Gelder, B. F.**, Demonstration by epr spectroscopy of the functional role of iron in soybean lipoxygenase, *Biochim. Biophys. Acta*, 377, 71, 1975.
115. **Holden, M.**, Lipoxidase activity of leaves, *Phytochemistry*, 9, 507, 1970.
116. **Matthew, J. A. and Galliard, T.**, Enzymic formation of carbonyls from linoleic acid in leaves of *Phaseolus vulgaris*, *Phytochemistry*, 17, 1043, 1978.
117. **Wardale, D. A., Lambert, E. A., and Galliard, T.**, Localisation of fatty acid hydroperoxide cleavage activity in membranes of tomato fruit, *Phytochemistry*, 17, 205, 1978.
118. **Sekiya, J., Kajiwara, T., and Hatanaka, A.**, Effect of inhibitors on the enzyme system producing C_6-aldehydes from C_{18} — unsaturated fatty acids in chloroplasts of Japanese silver (*Farfugium japonicum*) leaves, *Plant Cell Physiol.*, 19, 553, 1978.
119. **Hatanaka, A., Kajiwara, T., Sekiya, J., and Kido, Y.**, Formation of 12-oxo-*trans*-10-dodecenoic acid in chloroplasts from *Thea sinensis* leaves, *Phytochemistry*, 16, 1828, 1977.
120. **Hatanaka, A. and Harada, T.**, Formation of *cis*-3-hexenal, *trans*-2-hexenal and *cis*-3-hexenol in macerated *Thea sinensis* leaves, *Phytochemistry*, 12, 2341, 1973.
121. **Soda, K., Kido, T., and Asada, K.**, 2-Nitropropane dioxygenase from *Hansenula mrakii*. Generation and participation of superoxide ion in the reaction, in *Biochemical and Medical Aspects of Active Oxygen*, Hayaishi, O. and Asada, K., Eds., University of Tokyo Press, Tokyo, 1978, chap. 10.
122. **Hamilton, G. A., Adolf, P. K., de Jersey, J., DuBois, G. C., Dyrkacz, G. R., and Libby, R. D.**, Trivalent copper, superoxide and galactose oxidase, *J. Am. Chem. Soc.*, 100, 1899, 1978.
123. **McCord, J. M. and Fridovich, I.**, Superoxide dismutase. An enzymic function for erythrocuprein (haemocuprein), *J. Biol. Chem.*, 244, 6049, 1969.
124. **Fridovich, I.**, The biology of oxygen radicals, *Science*, 201, 875, 1978.
125. **Allen, J. F.**, A two-step mechanism for the photosynthetic reduction of O_2 by ferredoxin, *Biochem. Biophys. Res. Commun.*, 66, 36, 1975.
126. **Mehler, A. H.**, Studies on reactions of illuminated chloroplasts. Mechanism of the reduction of O_2 and other Hill reagents, *Arch. Biochem. Biophys.*, 33, 65, 1951.
127. **Asada, K., Kiso, K., and Yoshikawa, K.**, Univalent reduction of molecular O_2 by spinach chloroplasts on illumination, *J. Biol. Chem.*, 249, 2175, 1974.
128. **Harbour, J. R. and Bolton, J. R.**, Superoxide formation in spinach chloroplasts. Esr detection by spin-trapping, *Biochem. Biophys. Res. Commun.*, 64, 803, 1975.
129. **Miller, R. W. and Macdowall, F. D. H.**, Tiron free radical as a sensitive indicator of chloroplast photo-autoxidation, *Biochim. Biophys. Acta*, 387, 176, 1975.
130. **Lien, S. and San Pietro, A.**, On the reactivity of O_2 with photosystem I electron acceptors, *FEBS Lett.*, 99, 189, 1979.
131. **Jennings, R. C. and Forti, G.**, Involvement of O_2 during photosynthetic induction, in *Proc. 3rd Int. Congr. Photosynthesis*, Avron, M., Ed., Elsevier, Amsterdam, 1975, 735.
132. **Elstner, E. F., Wildner, G. F., and Heupel, A.**, Oxygen activation by isolated chloroplasts from *Euglena gracilis*. Ferredoxin-dependent function of a fluorescent compound and photosynthetic electron transport close to photosystem I, *Arch. Biochem. Biophys.*, 173, 623, 1976.
133. **Elstner, E. F., Stoffer, C., and Heupel, A.**, Determination of superoxide free radical ion and H_2O_2 as products of photosynthetic oxygen reduction, *Z. Naturforsch.*, 30C, 53, 1975.
134. **Trebst, A.**, Energy conservation in photosynthetic electron transport of chloroplasts, *Annu. Rev. Plant Physiol.*, 25, 423, 1974.
135. **Fuller, R. C. and Nugent, N. A.**, Pteridines and the function of the photosynthetic reaction centre, *Proc. Natl. Acad. Sci. U.S.A.*, 63, 1311, 1969.
136. **Iwai, K., Bunno, M., Kobashi, M., and Suzuki, T.**, Isolation and characterisation of 6-hydroxymethylpterin as the *Crithidia* growth-promoting factor from spinach chloroplasts, *Biochim. Biophys. Acta*, 444, 618, 1976.
137. **Nishikimi, M.**, Generation of superoxide anion in the reaction of tetrahydropteridines with molecular O_2, *Arch. Biochem. Biophys.*, 166, 273, 1975.
138. **Rich, P. R. and Lamb, C. J.**, Biophysical and enzymological studies upon the interaction of *trans*-cinnamic acid with higher plant microsomal cytochromes P450, *Eur. J. Biochem.*, 72, 353, 1977.
139. **Marchant, R. H.**, An analysis of the effect of added thiols on the rate of O_2 uptake during Mehler reactions which involve superoxide production, in *Proc. 3rd Int. Congr. Photosynthesis*, Avron, M., Ed., Elsevier, Amsterdam, 1975, 637.
140. **Bielski, B. H. J. and Chan, P. C.**, Products of reaction of superoxide and hydroxyl radicals with Mn^{2+} cation, *J. Am. Chem. Soc.*, 100, 1920, 1978.
141. **Fee, J. A.**, On the question of superoxide toxicity and the biological function of superoxide dismutases, in *Proc. 3rd Int. Congr. Oxidases and Related Redox Systems*, King, T. E., Mason, H. S., and Morrison, M., Eds., Pergamon Press, Oxford, 1982.

142. **Halliwell, B.**, Free radicals, oxygen toxicity and ageing, in *Age Pigments*, Sohal, R. S., Ed., Elsevier, Amsterdam, 1981, 1.
143. **Brown, R. C.**, Mechanism of iron uptake by plants, *Plant, Cell Environ.*, 1, 249, 1978.
144. **Elstner, E. F., Saran, M., Bors, W., and Lengfelder, E.**, O_2 activation in isolated chloroplasts. Mechanism of ferredoxin-dependent ethylene formation from methionine, *Eur. J. Biochem.*, 89, 61, 1978.
145. **Harbour, J. R. and Bolton, J. R.**, Involvement of the hydroxyl radical in the destructive photooxidation of chlorophylls *in vivo* and *in vitro*, *Photochem. Photobiol.*, 28, 231, 1978.
146. **Walling, C.**, Nature of the primary oxidants in oxidations mediated by metal ions, in *Proc. 3rd Int. Symp. Oxidases and Related Redox Systems*, King, T. E., Mason, H. S., and Morrison, M., Eds., Pergamon Press, Oxford, 1982.
147. **Rabinowitch, H. D. and Sklan, D.**, Superoxide dismutase. A possible protective agent against sunscald in tomatoes (*Lycopersicon esculentum* Mill.), *Planta (Berlin)*, 148, 162, 1980.
148. **Salin, M. L. and Bridges, S. M.**, Isolation and characterisation of an iron-containing superoxide dismutase from a eukaryote, *Brassica campestris*, *Arch. Biochem. Biophys.*, 201, 369, 1980.
149. **Jackson, C., Dench, J., Moore, A. L., Halliwell, B., Foyer, C. H., and Hall, D. O.**, Subcellular localisation and identification of superoxide dismutase in the leaves of higher plants, *Eur. J. Biochem.*, 91, 339, 1978.
150. **Asada, K., Urano, M., and Takahashi, M.**, Subcellular location of superoxide dismutase in spinach leaves and preparation and properties of crystalline spinach superoxide dismutase, *Eur. J. Biochem.*, 36, 257, 1973.
151. **Allen, J. F. and Hall, D. O.**, Superoxide reduction as a mechanism of ascorbate-stimulated O_2 uptake by isolated chloroplasts, *Biochem. Biophys. Res. Commun.*, 52, 856, 1973.
152. **Halliwell, B.**, Hydroxylation of *p*-coumaric acid by illuminated chloroplasts. The role of superoxide, *Eur. J. Biochem.*, 55, 355, 1975.
153. **Elstner, E. F. and Heupel, A.**, Lamellar superoxide dismutase of isolated chloroplasts, *Planta (Berlin)*, 123, 145, 1975.
154. **Doll, S., Lutz, C., and Ruppel, H. G.**, Biochemische und cytologische unkrsuchungen zur chloroplasten-entwicklung. Superoxiddismutase in etioplasten von *Avena sativa*, *Z. Pflanzenphysiol.*, 80, 166, 1976.
155. **Foyer, C. H. and Hall, D. O.**, A rapid procedure for the preparation of light-harvesting chlorophyll a/b protein complex, *FEBS Lett.*, 101, 324, 1979.
156. **Saetre, R. and Rabenstein, D. L.**, Determination of cysteine and glutathione in fruit by high-performance liquid chromatography, *Agric. Food Chem.*, 26, 981, 1978.
157. **Gowland-Hopkins, F. and Morgan, E. J.**, Appearance of glutathione during the early stages of germination of seeds, *Nature (London)*, 152, 288, 1943.
158. **Mapson, L. W. and Moustafa, E. M.**, Ascorbic acid and glutathione as respiratory carriers in the respiration of pea seedlings, *Biochem. J.*, 62, 248, 1956.
159. **Anderson, D. G., Stafford, H. A., Conn, E. E., and Vennesland, B.**, Distribution in higher plants of NADP-linked enzyme systems capable of reducing glutathione, *Plant Physiol.*, 27, 675, 1953.
160. **Drumm, H. and Mohr, H.**, Control by phytochrome of glutathione reductase levels in the mustard seedling, *Z. Naturforsch.*, 28C, 559, 1973.
161. **Uotila, L. and Koivusalo, M.**, Purification of formaldehyde and formate dehydrogenases from pea seeds by affinity chromatography and S-formylglutathione as the intermediate of formaldehyde metabolism, *Arch. Biochem. and Biophys.*, 196, 33, 1979.
162. **Fahey, R. C., Brody, S., and Mikolajczyk, S. D.**, Changes in the glutathione thiol-disulphide status of *Neurospora crassa* conidia during germination and aging, *J. Bacteriol.*, 121, 144, 1975.
163. **Corin, R. E. and Cox, C. D.**, Characterisation of leptospiral catalase and peroxidase, *Can. J. Microbiol.*, 26, 121, 1980.
164. **Omran, R. G.**, Peroxide levels and the activities of catalase, peroxidase and indoleacetic acid oxidase during and after chilling cucumber seedlings, *Plant Physiol.*, 65, 407, 1980.
165. **Soll, J., Douce, R., and Schultz, G.**, Site of biosynthesis of alpha-tocopherol in spinach chloroplasts, *FEBS Lett.*, 112, 243, 1980.
166. **Kornbrust, D. J. and Mavis, R. D.**, Microsomal lipid peroxidation. Characterisation of the role of iron and NADPH, *Mol. Pharmacol.*, 17, 400, 1980.
167. **Marsho, T. V., Behrens, P. W., and Radmer, R. J.**, Photosynthetic O_2 reduction in isolated intact chloroplasts and cells from spinach, *Plant Physiol.*, 64, 656, 1979.
168. **Nishikimi, M., Yamada, H., and Yagi, K.**, Oxidation by superoxide of tocopherols dispersed in aqueous media with deoxycholate, *Biochim. Biophys. Acta*, 627, 101, 1980.
169. **Sagone, A. L., Decker, M. A., Wells, R. M., and Democko, C.**, A new method for the detection of hydroxyl radical production by phagocytic cells, *Biochim. Biophys. Acta*, 628, 90, 1980.
170. **Misra, H. P. and Fridovich, I.**, Purification and properties of superoxide dismutase from a red alga *Porphyridium cruentum*, *J. Biol. Chem.*, 252, 6421, 1977.

171. **Lavelle, F. and Michelson, A. M.**, Purification et étude des deux superoxide dismutases du champignon *Pleurotus olearius, Biochimie,* 57, 375, 1975.
172. **Asada, K., Yoshikawa, K., Takahashi, M., Maeda, Y., and Enmanji, K.**, Superoxide dismutases from a blue-green alga *Plectonema boryanum, J. Biol. Chem.,* 250, 2801, 1975.
173. **Kano, Y., Takahashi, M., and Asada, K.**, Superoxide dismutases from kidney-bean leaves, *Plant Cell Physiol.,* 20, 1229, 1979.
174. **Okada, S., Kanematsu, S., and Asada, K.**, Intracellular distribution of manganic and ferric superoxide dismutases in blue-green algae, *FEBS Lett.,* 103, 106, 1979.
175. **Lumsden, J. and Hall, D. O.**, Soluble and membrane-bound superoxide dismutases in a blue-green alga (*Spirulina*) and spinach, *Biochem. Biophys. Res. Commun.,* 58, 35, 1974.
176. **Misra, H. P. and Keele, B. B.**, Purification and properties of superoxide dismutase from a blue-green alga, *Biochim. Biophys. Acta,* 379, 418, 1975.
177. **Lengfelder, E. and Elstner, E. F.**, Cyanide-sensitive iron superoxide dismutase in *Euglena gracilis.* Comparison of the reliabilities of different test systems for superoxide dismutases, *Z. Natuforsch.,* 34C, 374, 1979.
178. **Kanematsu, S. and Asada, K.**, Ferric and manganic superoxide dismutases in *Euglena gracilis, Arch. Biochem. Biophys.,* 195, 535, 1979.
179. **Henry, L. E. A. and Hall, D. O.**, Superoxide dismutases in green algae. An evolutionary survey, *Plant Cell Physiol.,* special issue on Photosynthetic Organelles, 377, 1977.
180. **Asada, K., Kanematsu, S., and Uchida, K.**, Superoxide dismutase in photosynthetic organisms. Absence of the cuprozinc enzyme in eukaryotic algae, *Arch. Biochem. Biophys.,* 179, 243, 1977.
181. **Giannopolitis, C. N. and Ries, S. K.**, Superoxide dismutases. Purification and quantitative relationship with water-soluble protein in seedlings, *Plant Physiol.,* 59, 315, 1977.
182. **Beauchamp, C. O. and Fridovich, I.**, Isozymes of superoxide dismutase from wheat-germ, *Biochim. Biophys. Acta,* 317, 50, 1973.
183. **Weser, U., Fretzdorff, A., and Prinz, R.**, Superoxide dismutase in bakers yeast, *FEBS Lett.,* 27, 267, 1972.
184. **Sawada, Y., Ohyama, T., and Yamazaki, I.**, Preparation and physicochemical properties of green pea superoxide dismutase, *Biochim. Biophys. Acta,* 268, 305, 1972.
185. **Baker, J. E.**, Superoxide dismutase in ripening fruits, *Plant Physiol.,* 58, 644, 1976.
186. **Giannopolitis, C. N. and Ries, S. K.**, Superoxide dismutases. Occurrence in higher plants, *Plant Physiol.,* 59, 309, 1977.
187. **Henry, L. E. A., Cammack, R., Schwitzguebel, J. P., Palmer J. M., and Hall, D. O.**, Intracellular localisation, isolation and characterisation of two distinct varieties of superoxide dismutase from *Neurospora crassa, Biochem. J.,* 187, 321, 1980.
188. **Baum, J. A. and Scandalios, J. G.**, Developmental expression and intracellular localisation of superoxide dismutases in maize, *Differentiation,* 13, 133, 1979.
189. **Ravindranath, S. D. and Fridovich, I.**, Isolation and characterisation of a manganese-containing superoxide dismutase from yeast, *J. Biol. Chem.,* 250, 6107, 1975.
190. **Hodgson, E. K. and Fridovich, I.**, Interaction of bovine erythrocyte superoxide dismutase with H_2O_2. Inactivation of the enzyme, *Biochemistry,* 14, 5294, 1975.
191. **Nishikimi, M.**, Oxidation of ascorbic acid with superoxide ion generated by the xanthine-xanthine oxidase system, *Biochem. Biophys. Res. Commun.,* 63, 463, 1975.
192. **Mees, G.**, Experiments on the herbicidal action of 1,1′-ethylene-2-2′dipyridylium dibromide, *Ann. Appl. Biol.,* 48, 601, 1960.
193. **Farrington, J. A., Ebert, M., Land, E. J., and Fletcher, K.**, Bipyridilium quaternary salts and related compounds. Pulse radiolysis studies of the reaction of paraquat radical with O_2. Implications for the mode of action of bipyridyl herbicides, *Biochim. Biophys. Acta,* 314, 372, 1973.
194. **Whitehouse, D. G., Ludwig, L. J., and Walker, D. A.**, Participation of the Mehler reaction and catalase in the O_2 exchange of chloroplast preparations, *J. Exp. Bot.,* 22, 772, 1971.
195. **Dodge, A. D.**, Some mechanisms of herbicide action, *Sci. Prog. (Oxford),* 62, 447, 1975.
196. **Harvey, B. M. R. and Fraser, T. W.**, Paraquat tolerant and susceptible perennial ryegrasses. Effects of paraquat treatment on CO_2 uptake and ultrastructure of photosynthetic cells, *Plant, Cell Environ.,* 3, 107, 1980.
197. **Takahama, U. and Nishimura, M.**, Effects of electron donor and acceptors, electron transfer mediators and superoxide dismutase on lipid peroxidation in illuminated chloroplast fragments, *Plant Cell Physiol.,* 17, 111 1976.
198. **Youngman, R. J., Dodge, A. D., Lengfelder, E., and Elstner, E. F.**, Inhibition of paraquat cytotoxicity by a novel copper chelate with superoxide dismutating activity, *Experientia,* 35, 1295, 1979.
199. **Harper, D. B. and Harvey, B. M. R.**, Mechanism of paraquat tolerance in perennial ryegrass. Role of superoxide dismutase, catalase and peroxidase, *Plant, Cell Environ.,* 1, 211, 1978.

200. **Krause, G. H., Kirk, M., Heber, U., and Osmond, C. B.**, O_2-dependent inhibition of photosynthetic capacity in intact isolated chloroplasts and isolated cells from spinach leaves illuminated in the absence of CO_2, *Planta (Berlin)*, 142, 229, 1978.
201. **Brennan, R. and Frenkel, C.**, Involvement of H_2O_2 in the regulation of senescence in pear, *Plant Physiol.*, 59, 411, 1977.
202. **Maguire, Y. P. and Haard, N. F.**, Fluorescent product accumulation in ripening fruit, *Nature (London)*, 258, 599, 1975.
203. **Cseke, C., Horvath, L. I., Simon, P., Borbely, G., Keozthelyi, L., and Farkas, G. L.**, An iron-containing superoxide dismutase from *Anacystis nidulans*, *J. Biochem. (Tokyo)*, 85, 1397, 1979.
204. **Rapp, U., Adams, W. C., and Miller, R. W.**, Purification of superoxide dismutase from fungi and characterisation of the reaction of the enzyme with catechols by esr spectroscopy, *Can. J. Biochem.*, 51, 158, 1973.
205. **Shatzman, A. R. and Kosman, D. J.**, Biosynthesis and cellular distribution of the two superoxide dismutases of *Dactylium dendroides*, *J. Bacteriol.*, 137, 313, 1979.
206. **Jursinic, P. A.**, Flash polarographic detection of superoxide production as a means of monitoring electron flow between photosystems I and II, *FEBS Lett.*, 90, 15, 1978.
207. **Quebedeaux, B. and Hardy, R. W. F.**, Reproductive growth and dry matter production of *Glycine max.* (L) Merr. in response to O_2 concentration, *Plant Physiol.*, 55, 102, 1975.
208. **Chin, C. K. and Frenkel, C.**, Upsurge in respiration and peroxide formation in potato tubers as influenced by ethylene, propylene and cyanide, *Plant Physiol.*, 59, 515, 1977.
209. **Henry, L. E. A., Gogotov, I. N., and Hall, D. O.**, Superoxide dismutase and catalase in the protection of the proton-donating systems of nitrogen fixation in the blue-green alga *Anabaena cylindrica*, *Biochem. J.*, 174, 373, 1978.
210. **Boeck, D. and Grosch, W.**, Glutathione-dehydrogenase (ECl.1.8.5) aus Weizenmehl. Reindarstellung und Eigenschaften, *Z. Lebensmittel Untersuchung Forsch.*, 162, 243, 1976.
211. **Yamaguchi, M. and Joslyn, M. A.**, Purification and properties of dehydroascorbic acid reductase of peas *(Pisum sativum)*, *Arch. Biochem. Biophys.*, 38, 451, 1952.
212. **Gero, E. and Candido, A.**, Purification de la reductase dehydroascorbique du chou-fleur (*Brassica oleracea* L.), *Bull. Soc. Chim. Biol.*, 49, 1895, 1967.
213. **Pulich, W. M.**, Resistance to high O_2 tension, streptonigrin and UV-irradiation in the green alga *Chlorella sorokiniana* strain ORS, *J. Cell Biol.*, 62, 904, 1974.
214. **Misra, H. P. and Fridovich, I.**, Generation of superoxide radical during the autoxidation of ferredoxins, *J. Biol. Chem.*, 246, 6886, 1971.
215. **Smith, J. and Shrift, A.**, Phylogenetic distribution of glutathione peroxidase, *Comp. Biochem. Physiol.*, 63B, 39, 1979.
216. **Foyer, C. H. and Hall, D. O.**, Oxygen metabolism in the active chloroplast, *Trends Biochem. Sci.*, 188, June 1980.
217. **Bridges, S. M. and Salin, M. L.**, Distribution of iron-containing superoxide dismutase in vascular plants, *Plant Physiol.*, 68, 275, 1981.

Chapter 6

FORMATION AND DISAPPEARANCE OF SUPEROXIDE RADICALS IN AQUEOUS SOLUTIONS

Augustine O. Allen and Benon H. J. Bielski

TABLE OF CONTENTS

I. HISTORY

The existence of the radical $HO_2\cdot$ in gas-phase reactions was first proposed by Marshall[1] in 1926 and has since been found to play a prominent role in atmospheric chemistry. Its existence in aqueous solution was first proposed by Haber and Willstatter[2] in 1931 as an intermediate in the decomposition of hydrogen peroxide by catalase, supposed by them (incorrectly, as we now know) to be a chain reaction, carried by the radicals $\cdot OH$ and $HO_2\cdot$:

$$\text{Cat} + H_2O_2 \longrightarrow \text{Cat-H} + HO_2\cdot \qquad (0)$$

$$HO_2\cdot + H_2O_2 \longrightarrow H_2O + O_2 + \cdot OH \qquad (1)$$

$$\cdot OH + H_2O_2 \longrightarrow H_2O + HO_2\cdot \qquad (2)$$

Haber and Weiss[3] applied this concept to explain the reaction between hydrogen peroxide and ferrous iron:

$$Fe^{++} + H_2O_2 \longrightarrow Fe^{+++} + OH^- + \cdot OH \qquad (3)$$

followed by Reactions 2 and 1 or by

$$\cdot OH + Fe^{++} \longrightarrow Fe^{+++} + OH^- \qquad (4)$$

The fact that at high peroxide concentrations the amounts of peroxide consumed and oxygen formed are much greater than the amount of iron oxidized were explained as being due to Reaction 1, which has become known as the "Haber-Weiss Reaction". Fenton[4] had shown long before that many organic acids and alcohols, ordinarily inert to peroxide, are rapidly oxidized by peroxide if ferrous salt is present. The formation of the very active oxidant $\cdot OH$ explains this effect.

In 1934, it was found[5] that the long-known crystalline superoxide of potassium, then called K_2O_4, was highly paramagnetic and must be regarded as composed of ions K^+ and $O_2^{\cdot-}$, the latter being paramagnetic by virtue of its free-radical character. Weiss[6] then pointed out that this ion in aqueous solution would be the basic form of $HO_2\cdot$, and postulated different reactivities of the two forms to explain effects of pH on rates of certain peroxide reactions. KO_2 had long been known[7] to react with water to form equal quantities of oxygen and peroxide, as one would expect from the disproportionation of $O_2^{\cdot-}$ radicals:

$$O_2^{\cdot-} + O_2^{\cdot-} + H_2O \longrightarrow HO_2^- + O_2 + OH^- \qquad (5)$$

where HO_2^- is the alkaline form of hydrogen peroxide. Bray[8] in 1938 pointed out that if Reaction 1 was occurring, one would expect more oxygen than peroxide, and Reaction 5 must be the only important reaction of $O_2^{\cdot-}$, and later experiments by Taube and Bray[9] verified this conclusion. Also photochemical studies on H_2O_2[10] were inconsistent with an important role for Reaction 1. In 1957, in a classical study in the field of reaction kinetics in solution by Barb et al.,[11] the oxidation of ferrous and reduction of ferric ions by H_2O_2 was carefully studied. The formation of oxygen gas was conclusively shown to result from the reduction of ferric ions by $HO_2\cdot$ or H_2O_2, so that Reaction 1 does not occur in any of the systems for which it was postulated. The subject is further discussed in a later section.

The development of nuclear energy during World War II required understanding of the effect of high-energy radiations on water and aqueous solutions. Water is decomposed to yield free radicals $\cdot H$ and $\cdot OH$, solvated electrons, and also molecules of hydrogen peroxide, which react with the radicals to yield $HO_2\cdot$ or $O_2^{\overline{\cdot}}$.[12] However, until 1961 the existence of the radical in aqueous solutions was only inferred from reaction kinetics. Development of the method of ESR spectroscopy allowed Saito and Bielski[13] to obtain in that year independent confirmation of the existence of $HO_2\cdot$. Optical spectra of $HO_2\cdot$ were obtained in 1962 by flash photolysis[14] and in 1964 by pulse radiolysis,[15] which in subsequent years has provided the most accurate data on the properties and behavior of $HO_2\cdot$ and $O_2^{\overline{\cdot}}$ in water solution.

The involvement of $O_2^{\overline{\cdot}}$ in biological processes was shown in 1958 by Fridovich and Handler,[16] who reported evidence that $O_2^{\overline{\cdot}}$ is formed when xanthine oxidase acts to oxidize xanthine or other substrates. The radical has since been shown to occur in many other enzymatic processes.

The latest chapter in the $O_2^{\overline{\cdot}}$ story concerns the formation of stable solutions of $O_2^{\overline{\cdot}}$ in aprotic solvents, i.e., liquids such as dimethyl sulfoxide (DMSO) which, though they may contain hydrogen, cannot donate protons to other substances. Electrolytic reduction of dissolved oxygen to $O_2^{\overline{\cdot}}$ in DMSO was first shown in 1965.[17] A more convenient preparation uses solutions in the aprotic solvent of a cyclic "crown" ether, which complexes potassium ions and hence solubilizes potassium salts in organic solvents. A 0.15 *M* solution of KO_2 with 0.3 *M* crown ether in DMSO remains stable for many days.[18]

II. METHODS OF FORMATION OF AQUEOUS $HO_2\cdot/O_2^{\overline{\cdot}}$

A. Radiolysis and Photolysis

Radiolysis of water produces free radicals $\cdot OH$ and $H\cdot$ and solvated electrons e_{aq}^-. If the water contains dissolved oxygen, $O_2^{\overline{\cdot}}$ and $HO_2\cdot$ are formed:

$$e_{aq}^- + O_2 \longrightarrow O_2^{\overline{\cdot}} \qquad (6)$$

$$H\cdot + O_2 \longrightarrow HO_2\cdot \qquad (7)$$

In the absence of a reducing agent, the $HO_2\cdot/O_2^{\overline{\cdot}}$ will be destroyed by the $\cdot OH$:

$$\cdot OH + HO_2\cdot \longrightarrow H_2O + O_2 \qquad (8)$$

If however the water contains dissolved hydrogen peroxide, all three of the primary radicals are converted to $HO_2\cdot/O_2^{\overline{\cdot}}$:

$$e_{aq}^- + H_2O_2 \longrightarrow \cdot OH + OH^- \qquad (9)$$

$$H\cdot + H_2O_2 \longrightarrow \cdot OH + H_2O \qquad (10)$$

$$\cdot OH + H_2O_2 \longrightarrow HO_2\cdot + H_2O \qquad (11)$$

Whether the product appears in the form of $HO_2\cdot$ or $O_2^{\overline{\cdot}}$ depends only on the pH of the solution, since the equilibrium $HO_2\cdot \rightleftharpoons H^+ + O_2^{\overline{\cdot}}$ is attained in aqueous solution more rapidly than any other reaction of the radicals.

Since peroxide, especially in alkaline solution, can behave quite similarly to $HO_2\cdot/O_2^{\cdot-}$, studies on properties of this radical are better done in the absence of H_2O_2. A system has been found[19] which converts all primary radicals of water radiolysis to $HO_2\cdot/O_2^{\cdot-}$ without using peroxide; it employs formate solutions containing oxygen. Reactions 6 and 7 convert e_{aq}^- and $\cdot H$, while the $\cdot OH$ reacts with formate:

$$\cdot OH + HCOO^- \longrightarrow H_2O + CO_2^- \qquad (12)$$

$$CO_2^- + O_2 \longrightarrow O_2^{\cdot-} + CO_2 \qquad (13)$$

Radiolysis may be produced by X-rays or gamma rays, but much higher intensities are given by beams of high-energy electrons produced by machines. In the technique known as pulse radiolysis, an accelerator is induced to inject into a solution a pulse of electrons which, though it may last only a fraction of a microsecond, carries enough energy to produce chemical changes. The absorption spectra and lifetimes of transient species such as $HO_2\cdot$ are then followed optically by fast electronics. Once the apparatus is set up, properties and behavior of transient species (previously postulated only by inference from masses of laboriously obtained data on kinetics of overall chemical changes) can be quickly and accurately determined. Thus the turnover rate for superoxide dismutase (SOD) has been determined far more accurately by pulse radiolysis[20] than is possible by other methods.

Photolysis of hydrogen peroxide solutions by UV light of wavelength 3800 to 2000 Å results in splitting the H_2O_2 molecule into two $\cdot OH$ radicals, which then yield $HO_2\cdot/O_2^{\cdot-}$ by Reaction 11. This method has not been used for preparation of $O_2^{\cdot-}$, because the high concentration of peroxide would interfere with study of $O_2^{\cdot-}$ effects. However, light of shorter wavelength (less than 1800 Å) is intensely absorbed by water and produces $\cdot H$ and $\cdot OH$ radicals in good yields, which in oxygenated 0.02 *M* formate solutions lead to $O_2^{\cdot-}$ formation.[21] Inexpensive argon- or xenon-filled lamps made with tubing of specially purified silica (Supersil) emit intensely in the appropriate wavelength range and provide a useful alternative to ionizing radiation for free-radical studies.

At acid or neutral pH, the lifetime of $HO_2\cdot/O_2^{\cdot-}$ is so short that it must be studied by pulse or fast-flow systems; but at high pH, its decay becomes slower (Figure 1). At pH 13 or above, solutions of 0.3 m*M* concentrations can readily be made by photolysis, X-rays, or gamma rays. They will decay very little over several minutes after removal from the radiation and can be studied by conventional methods.

B. Electrolysis

Stable solutions of $O_2^{\cdot-}$ are relatively easily prepared by electrolysis in liquid ammonia and in aprotic organic solvents,[17,22] but in water the reduction of oxygen normally goes all the way to H_2O_2 or even to H_2O. However, in alkaline solutions of surface-active materials, O_2 can be reduced to $O_2^{\cdot-}$,[23,24] which can be built up to concentrations as high as 1 m*M*.[25] It is thought that in the absence of surfactant, the $O_2^{\cdot-}$ formed at the electrode surface remains there long enough to pick up a second electron from the cathode, together with a proton from the water to form hydrogen peroxide anion:

$$H_2O + O_2^{\cdot-} + e^- \longrightarrow HO_2^- + OH^- \qquad (14)$$

In surfactant solutions the metal electrode adsorbs the nonpolar hydrocarbon part of the molecule and becomes covered with a hydrocarbon layer. When a negative

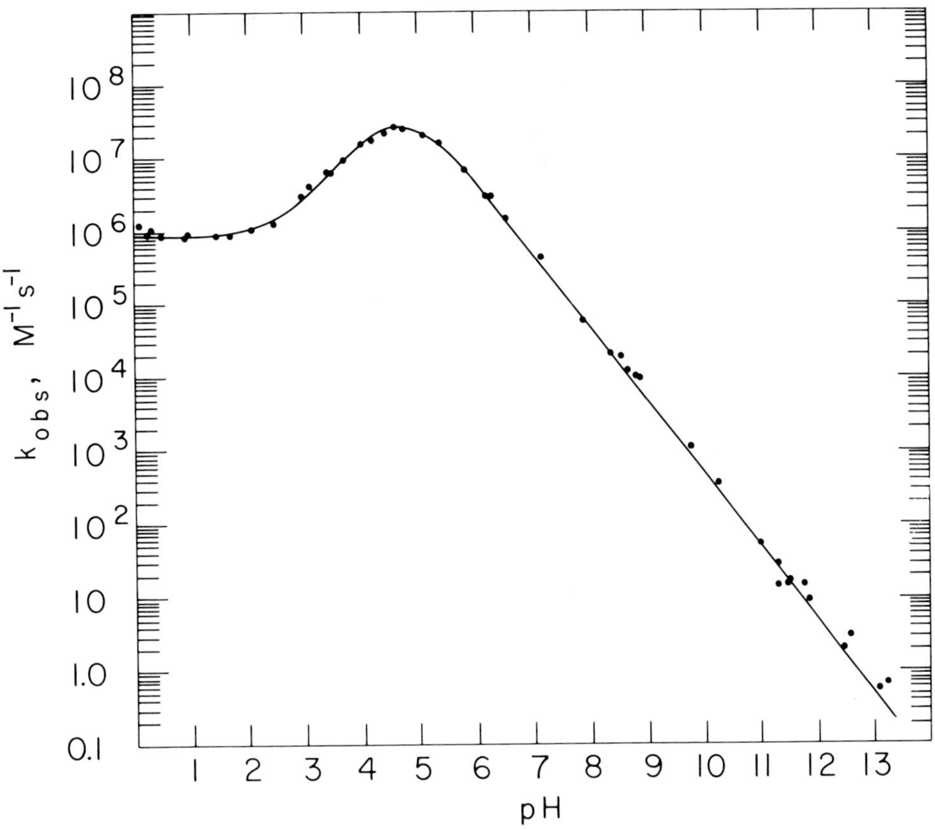

FIGURE 1. Rate constant for spontaneous second-order decay of $HO_2\cdot/O_2^{\bar{\cdot}}$.[55]

charge is applied to the metal, positive charge is induced on the polar ends of the surfactant layer. Oxygen molecules diffuse through this layer and accept an electron from the metal. The resulting $O_2^{\bar{\cdot}}$ ions tend to be pulled away from the metal by the induced positive charge on the outside of the layer and then become hydrated. The method seems to work better with a liquid mercury cathode than with solid metals. Successful $O_2^{\bar{\cdot}}$ synthesis in aqueous solution by this method requires more skill with electrochemical and polarographic methodologies than most chemists have.

C. Mixing Nonaqueous Solutions into Water

So-called "stop-flow" systems, in which flowing streams are mixed rapidly by mingling of high-speed jets and subsequent chemical reactions in the mixed stream are followed optically, have become widespread in recent years, with both streams usually being aqueous solutions. In such systems, well-designed mixers can achieve essentially complete mixing in 1 msec. With solvents of differing polarities, mixing is much slower, even under extreme turbulence. To mix DMSO solution of $O_2^{\bar{\cdot}}$ into water, the organic solution is passed into a large excess of water through tiny holes; then the resulting suspension is passed through two tangential mixers in series.[26] An optically homogeneous solution is obtained after 10 msec, allowing kinetic studies to be made of fast reactions of $O_2^{\bar{\cdot}}$ in the aqueous phase.

D. Chemical Reactions Producing $HO_2\cdot/O_2^{\bar{\cdot}}$

The radical may be formed chemically by reduction of oxygen or oxidation of hydrogen peroxide. If the reducing or oxidizing agent acts by transference of only one electron at a time, $HO_2\cdot/O_2^{\overline{\cdot}}$ will be present as a short-lived intermediate. It can be detected by adding a third reagent which is acted on by $HO_2\cdot/O_2^{\overline{\cdot}}$ but not by the reagents used for producing the radical, or more directly by carrying out the reaction in, or flowing the reaction mixture through, an ESR spectrometer. Because the ESR spectrum is sharper in frozen solutions, the reacting solution may be advantageously quickly frozen with liquid nitrogen, and the spectrum of the frozen solution then obtained.

Ceric sulfate, a powerful one-electron oxidizing agent in acid solution, rapidly oxidizes H_2O_2 to oxygen. When the reacting solution flowed through an ESR spectrometer, a spectrum typical of $HO_2\cdot$ was seen.[13] This method has been used to determine complex formation of $HO_2\cdot$ with a large number of metal ions.[27] In alkaline solutions, sodium periodate, $NaIO_4$, has been used with H_2O_2 to generate $O_2^{\overline{\cdot}}$.[28] There, the rapid freezing method was used to render the radical detectable by the ESR, since it is not seen by this method in water, nor in ice above −65°C.

In a related method, the Fenton reaction[4] may be used to generate $HO_2\cdot$. When ferrous or other strongly reducing metallic ions, such as titanous or vanadous, are mixed with excess hydrogen peroxide, Reaction 3 is quickly followed by Reaction 2, and complexes of the resulting $HO_2\cdot$ were identified by ESR when various metal ions were added to the mixture.[27]

A peroxide solution, made to flow over metallic platinum, palladium, or silver catalyst and then rapidly frozen showed the ESR spectrum of $O_2^{\overline{\cdot}}$; but the yield was much smaller than the amount of peroxide decomposed.[29]

A number of organic reducing agents can convert oxygen to $O_2^{\overline{\cdot}}$ without enzymatic help. Tetraacetyl riboflavin reacts with ethylenediamine tetraacetate (EDTA) under visible light to form the reduced flavin, the EDTA being oxidized to formaldehyde and other products. When this solution was mixed with oxygenated water in a flow system and rapidly frozen, the ESR spectrum of $O_2^{\overline{\cdot}}$ was seen; cytochrome *c* was reduced by the $O_2^{\overline{\cdot}}$, and the radical was destroyed by SOD.[30] The compounds 6-hydroxydopamine, 6-aminodopamine, 6,7-dihydroxytryptamine, and dialuric acid, dissolved in O_2-free water, were found to take up oxygen rapidly; cytochrome *c* added to the solution was reduced, indicating the presence of $O_2^{\overline{\cdot}}$,[31] while this reduction was inhibited if SOD was also added. In these reactions the $O_2^{\overline{\cdot}}$ was not isolated.

It is claimed[32] that in the alkaline decomposition of concentrated (10 to 30%) aqueous hydrogen peroxide, concentrations of $O_2^{\overline{\cdot}}$ approaching 2×10^{-3} *M* can be demonstrated by ESR with the fast-freezing method. Decomposition of H_2O_2 is fast when the pH equals the pK of H_2O_2 and the concentrations of H_2O_2 and HO_2^- are equal,[33,34] suggesting that the main reaction occurring is

$$H_2O_2 + HO_2^- \longrightarrow O_2 + H_2O + OH^- \tag{15}$$

$O_2^{\overline{\cdot}}$ may be a product of a side reaction, possibly

$$H_2O_2 + 2HO_2^- \longrightarrow 2O_2^{\overline{\cdot}} + 2H_2O \tag{16}$$

Presence of $O_2^{\overline{\cdot}}$ in concentrated alkaline peroxide could hardly be demonstrated any other way than by ESR, since the alkaline peroxide reacts similarly to $O_2^{\overline{\cdot}}$ with the usual indicators, and its own optical absorption will cover up that of $O_2^{\overline{\cdot}}$.

E. Enzymatic Generation of $O_2^{\overline{\cdot}}$

Of the many enzymes which apparently generate superoxide radicals during enzymatic activity, only a few are of practical use in the study of $O_2^{\overline{\cdot}}$. These are the oxidoreductases which generate free $O_2^{\overline{\cdot}}$, which can diffuse from the point of origin (active site) into the bulk of the solution for further reaction. The four proven enzymes to generate free $O_2^{\cdot-}$ are xanthine oxidase (xanthine: oxygen oxidoreductase 1.2.3 .2),[35-38] aldehyde oxidase (aldehyde: oxygen oxidoreductase 1.2.3.1),[39-42] dihydroorotate oxidase (L-5,6-dihydroorotate: oxygen oxidoreductase 1.3.3.1),[40,43-45] and NADPH-cytochrome reductase (NADPH: ferricytochrome oxidoreductase 1.6.2.4).[46,47] The enzyme most frequently used at present is xanthine oxidase because of its commercial availability in relatively pure form.

Xanthine oxidase is a very complex enzyme which contains 2 mol of flavin adenine dinucleotide (FAD), 8 mol of iron, 2 mol of molybdenum, and labile sulfide groups. Its catalytic specificity is low both with respect to substrate oxidation (hypoxanthine, xanthine, pyrimidines, aldehydes, pterins, reduced nicotinamide adenine dinucleotide [NADH], etc.) and electron acceptors (oxygen, ferricyanide, artificial dyes). Although this is one of the most extensively studied enzymes, how it transfers electrons from donor to acceptor is not yet well understood.

In the standard laboratory method[35] for superoxide radical generation the preferred substrates are either xanthine or acetaldehyde, which are oxidized to uric acid and acetic acid, respectively. In the enzymatic process reduced xanthine oxidase apparently loses electrons by two different pathways, a univalent reaction yielding $O_2^{\cdot-}$ (Reaction 17) and a divalent reaction which leads directly to hydrogen peroxide formation (Reaction 18):

$$\text{Enzyme-H}_2 + 2O_2 \longrightarrow \text{Enzyme} + 2H^+ + 2O_2^{\cdot-} \quad (17)$$

$$\text{Enzyme-H}_2 + O_2 \longrightarrow \text{Enzyme} + H_2O_2 \quad (18)$$

The percentage yield of superoxide radical in this system was found to vary from 20 to 100% depending upon pH, oxygen concentration, and the turnover rate of the enzyme. In general raising the pH and oxygen concentration or lowering the turnover rate favors Reaction 17.

Although the xanthine oxidase system is a very useful source of superoxide radicals for steady-state studies, great caution should be exercised in its use. The presence of hydrogen peroxide and inherent small amounts of metallic impurities in such enzyme systems can easily lead to Fenton-type reactions and hence to misinterpretation of observed results.

F. Photosensitization

Photosensitized oxidation has long been known. Many compounds, illuminated in the presence of air and a colored sensitizer (often a dye such as Rose Bengal [RB] or methylene blue) are oxidized or degraded. In the past, many such reactions have been attributed to the formation of energy-rich molecules of singlet oxygen $^1O_2$ by reaction of the light-excited sensitizer with ordinary triplet oxygen. It now seems that many such reactions involve $O_2^{\cdot-}$ instead of, or in addition to, singlet oxygen. For instance, the chain oxidation of sulfite by air is greatly speeded up by illumination in the presence of RB.[48] The photoreaction is partially but not completely inhibited by SOD, which destroys $O_2^{\cdot-}$, and also by 1,4-diazobicyclo[2.2.2]octane (DABCO), which is known to react rapidly with $^1O_2$. Thus the reaction is partially mediated by $O_2^{\cdot-}$ and also by another pathway involving $^1O_2$. Oxidation of another compound, 2,3-diphenyl-*p*-dioxene, was inhibited only by DABCO, not by SOD so that in that case no $O_2^{\cdot-}$ was involved.

A different type of sensitizer is represented by the bipyridyl-complexed divalent ruthenium ion, Ru $(bipy)_3^{2+}$. Illuminated in acetonitrile solution, this accepts an electron from any good electron donor (ED) which may be present to form the unstable reduced ion, Ru (bipy)$_3^+$, seen as a transient spectrum along with the spectrum of the ED^+ ion.[49] When oxygen was present, the spectrum of Ru (bipy)$_3^+$ did not appear, only ED^+, and it must be assumed that O_2 oxidizes Ru back to its original form, while taking on an electron to become $O_2^{\overline{\cdot}}$. The ion ED^+ disappears by a rapid second-order process, obviously $O_2^{\overline{\cdot}} + ED^+$, so no net chemical change occurs in the end.

More direct evidence for photosensitized $O_2^{\overline{\cdot}}$ formation is provided by the method of "spin trapping". Certain nitroxides, such as 5,5′-dimethyl-1-pyrroline-1-oxide (DMPO) readily add a free radical R· to form a very stable nitroxyl radical:

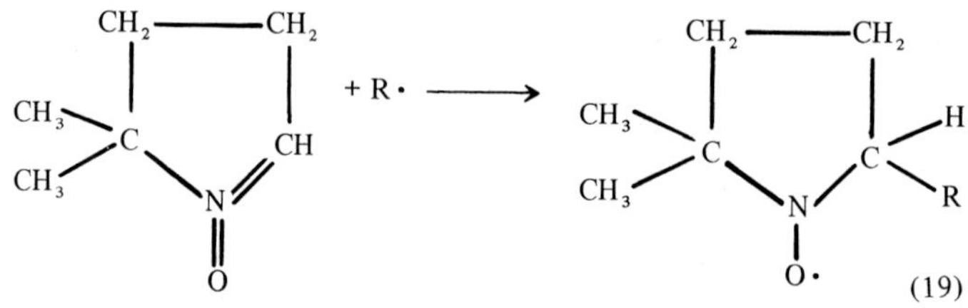

(19)

The ESR spectrum of the product radical then provides information on the nature of R·. The $O_2^{\overline{\cdot}}$ adduct of DMPO appears on illumination of its solution in aqueous H_2O_2.[50] By this method it was proved[51] that $O_2^{\overline{\cdot}}$ forms when a suspension in water or acetonitrile of a semiconductor, such as cadmium sulfide or phthalocyanine, is illuminated. Solutions of metalloporphyrins in aprotic solvents were also shown by spin trapping to give a good yield of $O_2^{\overline{\cdot}}$,[52] but in these cases there was also evidence for large yields of $^1O_2$.

One aromatic photosensitizing molecule, 9,10-dicyanoanthracene (DCA), has been reported[53] which appears to operate through $O_2^{\overline{\cdot}}$ exclusively, with essentially no yield of $^1O_2$. However, the existence of free $O_2^{\overline{\cdot}}$ was not established. The EDs used all quenched the DCA fluorescence very efficiently, and the initial step was thought to be complex formation with partial or complete charge transfer in the complex:

$$\mathrm{DCA} \xrightarrow{h\nu} \mathrm{DCA}^* \xrightarrow{\mathrm{ED}} [\mathrm{DCA}^- - \mathrm{ED}^+] \xrightarrow{O_2} \begin{bmatrix} \mathrm{DCA} - \mathrm{ED}^+ \\ O_2^{\overline{\cdot}} \end{bmatrix} \longrightarrow \mathrm{DCA} + \text{products} \qquad (20)$$

Thus the whole process could occur in the vicinity of the excited DCA molecule without the $O_2^{\overline{\cdot}}$ ever becoming free. The observed products included benzaldehyde from stilbene, benzophenone from tetraphenylene, and sulfoxides from diethyl and diphenyl sulfides, which could not be formed from $^1O_2$ in the observed yields.

III. PROPERTIES OF $HO_2\cdot/O_2^{\overline{\cdot}}$ IN AQUEOUS SOLUTION

A. Spontaneous Dismutation Rates, pK, and Absorption Spectra

The pulse radiolysis method is especially well suited to determining lifetimes of transient species such as $HO_2\cdot/O_2^{\overline{\cdot}}$. An early paper[54] showed that these species have a strong optical absorption in the UV, more intense in neutral than in acid solutions, which can be used to trace the course of the decay in time. In this work, the decay was mainly kinetically second order at all pHs studied, but also appeared to have a smaller first-order component, which could be due to catalysis by impurities. The study was then carried on at another laboratory, with improved pulse radiolysis techniques and sample purification.[19] Purely second-order decays were found from pH 0 to 8, but a slow first-order component, ascribed to impurities appeared in alkaline solutions. The best spectra of $HO_2\cdot$ and $O_2^{\overline{\cdot}}$ available at present are shown in Figure 2.[55] The spectrum seen near the pK is intermediate and depends on the ratio of the two components present, so that from the change in the spectra as pH is varied, the pK can be determined.

The decay rate, nearly constant at pH less than 2, rises to a maximum near the pK, where the two forms are of equal concentration, and then becomes much smaller in alkali (Figure 1). This behavior is explained by the necessity of having at least two different bimolecular disproportionation reactions:

$$HO_2\cdot + HO_2\cdot \longrightarrow H_2O_2 + O_2 \qquad (21)$$

$$HO_2\cdot + O_2^{\overline{\cdot}} \longrightarrow HO_2^- + O_2 \qquad (22)$$

HO_2^- is the basic form of hydrogen peroxide. The maximum in the curve of rate vs. pH occurs because Reaction 22 is faster than Reaction 21. Three constants are required to fit the curve, k_{21}, k_{22}, and the pK. Thus the data of these experiments provide two independent methods, kinetic and spectroscopic, for determination of the pK of $HO_2\cdot$.

The third possibility for a disproportionation reaction is, of course

$$O_2^{\overline{\cdot}} + O_2^{\overline{\cdot}} + H_2O \longrightarrow HO_2^- + O_2 + OH^- \qquad (5)$$

This reaction was shown[19,54] to be much slower than Reaction 21 or 22, but its rate could not be determined because of the occurrence of the impurity-catalyzed first-order disappearance of $O_2^{\overline{\cdot}}$. The work was therefore repeated[56] with specially purified reagents, and with the complexing agent EDTA present to sequester possibly catalytic metal ions. The result was that Reaction 5 does not occur at all ($k_5 < 0.3$ M^{-1} sec^{-1}) and a 10^{-4} *M* solution of $O_2^{\overline{\cdot}}$ at pH 13 has an initial half-life of about 160 min.

Recently[55] a careful redetermination of the optical absorption coefficients led to the spectrum shown in Figure 1 and the recalculated disproportionation rates plotted in Figure 2. The $pK_{HO_2\cdot}$ is 4.69 ± 0.08 (about 0.2 less than the often-quoted earlier value); $k_{21} = 8.6 \times 10^5$ M^{-1} sec^{-1} and $k_{22} = 1.02 \times 10^8$ M^{-1} sec^{-1}.

B. Catalyzed Dismutation

Especially in solutions above pH 7, the catalyzed dismutation is thus more important practically than the spontaneous reaction and has been extensively studied. The most famous catalysts are of course the SODs, which are proteins containing copper, manganese, or iron. The rate of the bimolecular reaction between $O_2^{\overline{\cdot}}$ and enzyme, leading to destruction of $O_2^{\overline{\cdot}}$, has been determined by pulse radiolysis to be 1.3×10^9 M^{-1} sec^{-1} per copper atom for bovine erythrocuprein[20,57,58] and coincidentally the same number for a manganese-containing enzyme from *E. coli.*[59] The erythrocuprein contains a zinc atom as well as copper, which can be replaced by cobalt; this replacement had no effect on the reaction with $O_2^{\overline{\cdot}}$.[60]

A large number of copper compounds have been tested for the rate at which they react in aqueous solution with $O_2^{\overline{\cdot}}$ to catalyze its dismutation; the results are shown in Table 1. It is noteworthy that plain uncomplexed aqueous copper ion is an even more effective "dismutase" than any enzyme. Soluble copper complexes may show catalytic activity, ranging unpredictably from very high to zero; the behavior towards $O_2^{\overline{\cdot}}$ depends on the detailed arrangement of ligands in a way which is not well understood. In the case of copper-histidine complexes, six different complexes have been characterized with different Cu/Hist ratios and degrees of protonation; of these, one ($CuHist_2H^{3+}$) reacts rapidly with $O_2^{\overline{\cdot}}$ ($k = 3.4 \times 10^8$ M^{-1} sec^{-1}), but the other five do not react noticeably at all.[61] One doubts that the complexes with all the other ligands listed in Table 1 have been as thoroughly characterized as those with histidine, and

Table 1
RATE CONSTANTS FOR REACTION OF $O_2^{\cdot-}$ WITH COPPER COMPOUNDS IN AQUEOUS SOLUTION LEADING TO CATALYTIC DISMUTATION OF $O_2^{\cdot-}$ [a]

Compound	pH	k, $M^{-1}\ s^{-1}$	Ref.
Cu_{aq}^{++}	2	2.7×10^9	58
Bovine erythrocuprein (SOD)	7—9	1.3×10^9	20,57,58
Cu_2 (Indomethacin)$_4$	7	6.0×10^9	62
Cu (Tyrosine)$_2$		1.3×10^9	60
Cu $(HCOO)^{+}$[b]	6	1.7×10^9	63
Cu $(HCOO)_2$	7	3.0×10^8	63
Cu $(HCOO)_3$	7	8.0×10^8	63
Cu $(HCOO)_4^{2-}$	8	4.0×10^8	63
Cu (NH_3) complexes	∿10	$0.2—2.2 \times 10^9$	63
$CuSO_4$	6	5.0×10^8	63
$CuCl_3^-$	7	7.0×10^8	63
Cu (Histidine-H)$^{3+}$		3.4×10^8	61
Cu (Lysine)$_2$		6.0×10^8	58
Cu (Glycine)$_2$	8	2.0×10^6	63
Cu (Glycylglycine)$_2$	7	2.0×10^7	63
Cu (Glycylhistidine)$_2$		3.0×10^8	58
Cu (Glycylhistidylleucine)$_2$		2.0×10^8	58
Cu (Alanine)$_2$	7	3.0×10^6	63
Cu-glutamate	7	1.0×10^6	63
Cu (Hydroxyproline)	8	1.0×10^6	63
Cu (Methionine)$_x$	7—8	$5.0—8.0 \times 10^6$	63

[a] Calculated on the basis of the number of copper atoms present.
[b] Ionic strength 2.0 was maintained in the formate studies. Values for zero ionic strength may differ by an order of magnitude.

the numbers given may in some cases be average values for whatever complexes may have been present under the particular conditions of the experiments.

Although it had been suggested in the past that the copper-catalyzed two-step dismutation might involve a copper-superoxide radical complex,[18,64,65] overall the mechanism was believed to cycle back and forth between Cu^{++} and Cu^+:

$$Cu^{++} + O_2^{\cdot-} \longrightarrow CuO_2^{\ +} \longrightarrow Cu^+ + O_2 \tag{23}$$

$$Cu^+ + O_2^{\cdot-} + 2H^+ \longrightarrow Cu^{++} + H_2O_2 \tag{24}$$

More recently[63] it was shown, by mixing a large excess of Cu^{++} with $HO_2\cdot/O_2^{\cdot-}$ ($Cu^{++}/HO_2\cdot\ 10^3-10^4$) that indeed one or more complexes are formed, which absorb strongly in the UV region:

$$Cu^{++} + O_2^{\cdot-} \longrightarrow CuO_2^{\ +} \tag{25}$$

$$CuO_2^{\ +} + Cu^{++} \longrightarrow (Cu^{++}\ldots O_2^{\cdot-}\ldots Cu^{++}) \tag{26}$$

This system is extremely complex since the complexes can be present in various states of protonation on which is superimposed the equilibrium $HO_2\cdot \rightleftharpoons O_2^{\cdot-} + H^+$.

As in the spontaneous disproportionation of superoxide radicals, where the control-

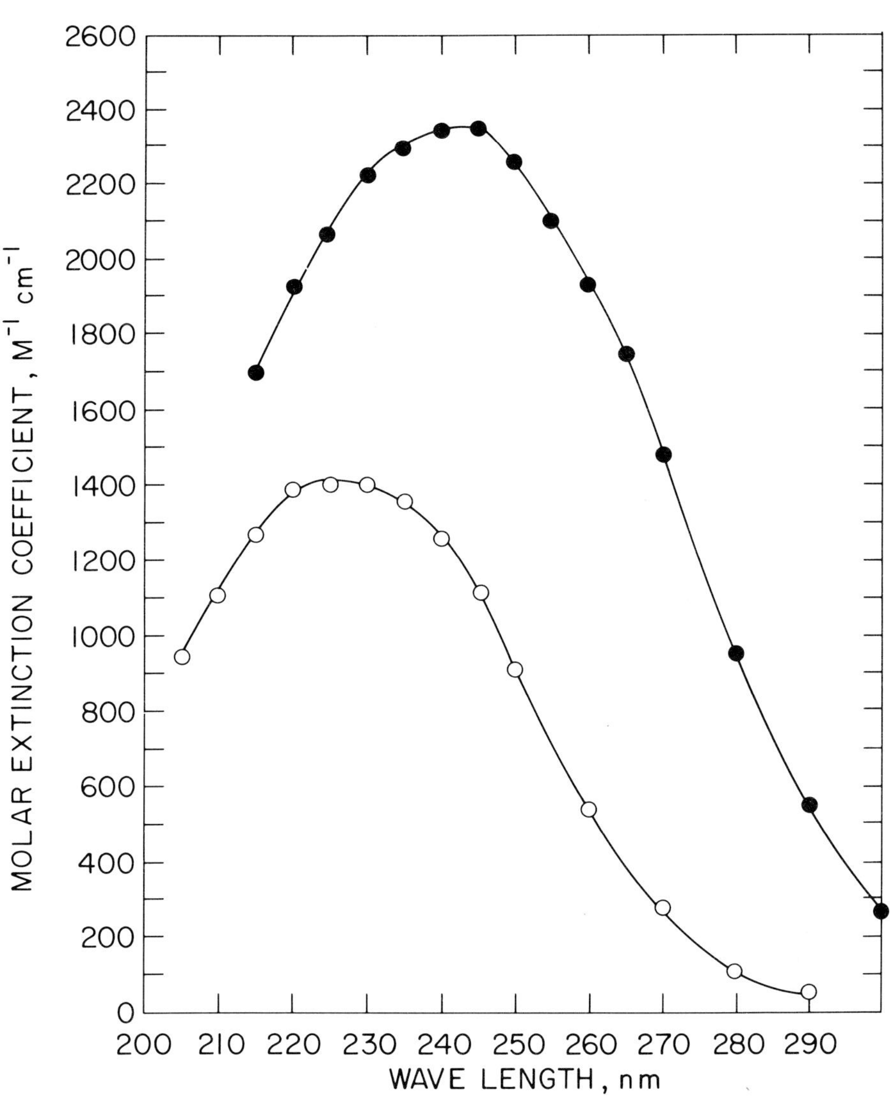

FIGURE 2. Absorption spectra of $O_2^{\cdot-}$ at pH 7 (upper curve) and $HO_2\cdot$ at pH 2 (lower curve).[55]

ling step is Reaction 22 between $HO_2\cdot$ and $O_2^{\cdot-}$, it was suggested[66] that in the copper catalyzed dismutation the Cu^{++}-cation plays the same role as the proton in $HO_2\cdot$;

$$Cu^{++} + O_2^{\cdot-} \longrightarrow CuO_2^{+} \qquad (25)$$

$$CuO_2^{+} + O_2^{\cdot-} + 2H_2O \longrightarrow Cu^{++} + H_2O_2 + O_2 + 2OH^{-} \qquad (27)$$

While complex formation between $O_2^{\cdot-}$ and metal cations (Fe^{++};[11,67] Mn^{++};[68,69] Ba^{++}, Ca^{++};[70] Ni^{++}, Cr^{++};[71,72] V^{5+}, Ti^{4+}, Ce^{3+}, Zr^{4+}, Th^{4+}, U^{6+}, Mo^{6+} [27]) is well documented, convincing evidence for a SOD-$O_2^{\cdot-}$ complex is lacking.

Although copper-EDTA complexes do not react with $O_2^{\cdot-}$, iron-EDTA complexes do. The reactions were studied[73] in a stopped-flow system in which a small volume of a DMSO solution of $O_2^{\cdot-}$ was mixed with a large volume of very dilute aqueous Fe^{++}-EDTA or Fe^{+++}-EDTA complex. The reaction formed the same purple peroxo complex

Table 2
STANDARD FREE ENERGIES OF FORMATION $\Delta_g(f)$ FOR VARIOUS SPECIES IN AQUEOUS SOLUTION AT 25°C

Compound	$\Delta_g(f)$, kcal/mol	Ref.
H_2O (liquid)	−56.687	76
OH^-	−37.594	76
H_2O_2	−32.05	76
HO_2^-	−16.1	76
$HO_2\cdot$	+1.2	25
$\cdot OH$	+4.5	78
$O_2^{\overline{\cdot}}$	+7.6	77

$[Fe^{+++}\text{-EDTA-}O_2]^{3-}$ that is formed from hydrogen peroxide and Fe^{++}-EDTA at high alkalinity. The redox reactions which occurred at pH 9.7 resulted in the disproportionation of O_2 :

$$[Fe^{++}\text{-EDTA}]^{2-} + O_2^{\overline{\cdot}} \longrightarrow [Fe^{+++}\text{-EDTA-}O_2]^{3-} \quad (28)$$

$$[Fe^{+++}\text{-EDTA-}O_2]^{3-} + 2H^+ \rightleftharpoons [Fe^{+++}\text{-EDTA}]^- + H_2O_2\cdot \quad (29)$$

$$[Fe^{+++}\text{-EDTA}]^- + O_2^{\overline{\cdot}} \longrightarrow [Fe^{++}\text{-EDTA}]^{2-} + O_2 \quad (30)$$

However, Reaction 29 requires seconds and the induced disproportionation rate is barely faster than the spontaneous disproportionation at this pH, so Fe-EDTA is not really much of a catalyst.

In contrast to the much-studied copper situation, we have found no work on the reactions of $O_2^{\overline{\cdot}}$ with other forms of iron in alkaline solution. In acid solution, the reactions of Fe^{++} and Fe^{+++} with $HO_2\cdot$ and H_2O_2 have been thoroughly worked out.[11,67] In biological systems, complexed ferrous iron may be important in converting hydrogen peroxide by the Fenton reaction to the very energy-rich oxidant $\cdot OH$.

Manganese, the other metal found in some SODs, has been relatively little studied with respect to its reactions with $O_2^{\overline{\cdot}}$. The usual form in aqueous solution, Mn^{++}, reacts with $O_2^{\overline{\cdot}}$ in the equilibrium reaction[68,69]

$$Mn^{++} + O_2^{\overline{\cdot}} \rightleftharpoons MnO_2^+ \quad (31)$$

This complex does not go to any higher oxidation state of Mn, and Mn_{aq}^{++} does not catalyze dismutation of $O_2^{\overline{\cdot}}$. However, in Mn-containing dismutases the metal is normally in the trivalent state, but can be reduced to the divalent state without loss of enzymatic activity.[74] Mn^{+++} is stable only in organic complexes. As a redox model for the action of the enzyme, complexes of Mn with 8-quinolinol were reacted with $O_2^{\overline{\cdot}}$ in DMSO solution.[75] A bewildering variety of complexes appeared to be involved in the reactions, but oxidation and reduction of the manganese surely occurred and dismutation of $O_2^{\overline{\cdot}}$ resulted.

C. Thermodynamics

The free-energy relationships of the stable compounds of hydrogen and oxygen have long been known very accurately.[76,77] To bring the radicals $HO_2\cdot$ and $O_2^{\overline{\cdot}}$ into this picture and establish their standard electrode potentials required a determination of the electrode potential of $O_2^{\overline{\cdot}}$, corresponding to the half-reaction

Table 3
STANDARD ELECTRODE POTENTIAL FOR SOME REDOX REACTIONS AT 25°C RELATIVE TO THE HYDROGEN ELECTRODE

Reaction	E° (volts)
$\cdot OH + e^- \rightarrow OH^-$	+ 1.83
$O_2^{\overline{\cdot}} + e^- + 2H^+ \rightarrow H_2O_2$	+ 1.73
$HO_2\cdot + e^- + H^+ \rightarrow H_2O_2$	+ 1.45
$O_2^{\overline{\cdot}} + e^- + H^+ \rightarrow HO_2^-$	+ 1.03
$O_2 + e^- + 2H^+ \rightarrow H_2O_2$	+ 0.695
$O_2 + 2e^- + H^+ \rightarrow HO_2^-$	+ 0.35
$e^- + H^+ \rightarrow \frac{1}{2}H_2$	0.000
$O_2 + e^- + H^+ \rightarrow HO_2\cdot$	−0.05
$O_2 + e^- \rightarrow O_2^{\overline{\cdot}}$	−0.33

Note: Standard states are, for H_2 and O_2, gas at 1 atm; H_2O (liquid); other substances, unit molality

$$O_2 + e^- \longrightarrow O_2^{\overline{\cdot}} \qquad (32)$$

Good electrochemical determinations have been made of this quantity,[25,78] but the most accurate method[79] measures the equilibrium between O_2, $O_2^{\overline{\cdot}}$, and quinone-semiquinone couple of known potential:

$$H^+ + O_2^{\overline{\cdot}} + Q \rightleftharpoons O_2 + QH\cdot \qquad (33)$$

From the equilibrium constant, the standard free-energy change is obtained. Standard states of most solutes are unit molality in water, but two usages can be found in the case of oxygen: its free energy can be related to a 1 *m* solution of the gas, or to the gas at atmospheric pressure, which corresponds to a 1.25×10^{-3} *M* aqueous solution. Here we use the latter system. In Table 2 we list the standard free energy of formation from the elements (H_2 and O_2) of various compounds in aqueous solution at 25°C. From the values in this table, together with the pKs of $HO_2\cdot$ (4.7) and H_2O_2 (11.6) we obtain the electrode potentials listed in Table 3.

D. The So-Called Haber-Weiss Reaction

There is no question that the great papers of Haber and Weiss[3] marked the opening of the whole field of free-radical chemistry in aqueous solution, and these men deserve to be known as the fathers of this field of chemistry. Of course, it is too much to expect even of these brilliant chemists that in the very beginning of the subject they should get every detail of their mechanisms right; as described at the beginning of this chapter, in their mechanisms for the iron-peroxide reaction, they proposed as a single step the reaction between $HO_2\cdot$ and H_2O_2

$$HO_2\cdot + H_2O_2 \longrightarrow H_2O + O_2 + \cdot OH \qquad (1)$$

which was later shown to occur in two steps

$$HO_2\cdot + Fe^{+++} \longrightarrow O_2 + Fe^{++} + H^+ \qquad (34)$$

$$Fe^{++} + H_2O_2 \longrightarrow Fe^{+++} + \cdot OH + OH^- \qquad (3)$$

Reaction 3 is of course the long-known Fenton reaction,[4] and this name is used to describe the generation of ·OH radicals from hydrogen peroxide by metal-containing reducing agents generally.

In more recent years, many biochemists have noticed that $O_2^{\cdot-}$ and H_2O_2, generated together enzymatically, led to oxidative effects suggesting formation of a powerful active oxidizing agent, reasonably assumed to be the ·OH radical. Not being aware of all the evidence amassed over the decades by physical chemists against Reaction 1, they attributed the formation of ·OH to this reaction and dubbed it "Haber-Weiss". By now the facts are obvious to all;[80] first, Reaction 1 does not occur spontaneously; second, these biological effects sugesting ·OH radical production occurred in the presence of metal-containing enzymes and should be attributed to reactions of the Fenton type.

Thus the controversy that has arisen is not over a matter of substance, but merely over a name. Especially those of us who knew and admired Joseph Weiss for his many fine achievements in the development of free-radical chemistry hate to see his name immortalized through attachment to an early proposal which proved to be a mistake. Surely, if the reaction generating ·OH radicals (or a similar strong oxidant) must be referred to by name, it should be given its proper designation, by which it was known long before Haber and Weiss published their theory, namely, the Fenton Reaction.

ACKNOWLEDGMENTS

Supported by the U.S. Department of Energy and by its Office of Basic Energy Sciences, and in part by National Institutes of Health Grant Number RO 1 GM 23656-05.

REFERENCES

1. **Marshall, A. L.,** Photosensitization by optically excited mercury atoms. The hydrogen-oxygen reaction, *J. Phys. Chem.*, 30, 44, 1926.
2. **Haber, F. and Willstätter, R.,** Unpaarigkeit und Radikalketten im Reaktionsmechanismus organischer und enzymatischer Vorgänge, *Ber. Deutsch. Chem. Ges.*, 64, 2844, 1931.
3. **Haber, F. and Weiss, J.,** Uber die Katalyse des Hydroperoxydes. *Naturwissenschaft*, 20, 948, 1932; The catalytic decomposition of hydrogen peroxide by iron salts, *Proc. R. Soc. (London)*, A147, 332, 1934.
4. **Fenton, H. J. H. and Jackson, H.,** The oxidation of polyhydric alcohols in presence of iron, *J. Chem. Soc. (London)*, 75, 1, 1899.
5. **Neuman, E. W.,** Potassium superoxide and the three-electron bond, *J. Chem. Phys.*, 2, 31, 1934.
6. **Weiss, J.,** Investigations on the radical HO_2 in solution, *Trans. Farad. Soc.*, 31, 668, 1935.
7. **Harcourt, A. V.,** On the peroxides of potassium and sodium, *J. Chem. Soc. (London)*, 14, 267, 1861.
8. **Bray, W. C.,** Interaction of ozone and hydrogen peroxide in aqueous solution, *J. Am. Chem. Soc.*, 60, 82, 1938.
9. **Taube, H. and Bray, W. C.,** Chain reactions in aqueous solutions containing ozone, hydrogen peroxide and acid, *J. Am. Chem. Soc.*, 62, 3357, 1940.
10. **Callow, A. E., Griffith, R. O., and McKeown, A.,** The photo-reaction between bromine and hydrogen peroxide in aqueous solution, *Trans. Farad. Soc.*, 35, 412, 1939.
11. **Barb, W. G., Baxendale, J. H., George, P., and Hargrave, K. R.,** Reactions of ferrous and ferric ions with hydrogen peroxide, *Trans. Farad. Soc.*, 47, 462 and 591, 1951.
12. **Allen, A. O.,** Radiation chemistry of aqueous solutions, *J. Phys. Chem.*, 52, 479, 1948.
13. **Saito, E. and Bielski, B. H. J.,** The electron paramagnetic resonance spectrum of the HO_2 radical in aqueous solution, *J. Am. Chem. Soc.*, 83, 4467, 1961.

14. **Baxendale, J. H.**, The flash photolysis of water and aqueous solutions, *Radiat. Res.*, 17, 312, 1962.
15. **Czapski, G. and Dorfman, L. M.**, Pulse radiolysis studies. V. Transient spectra and rate constants in oxygenated aqueous solutions, *J. Phys. Chem.*, 68, 1169, 1964.
16. **Fridovich, I. and Handler, P.**, Xanthine oxidase. IV. Participation of iron in internal electron transport, *J. Biol. Chem.*, 223, 1581, 1958.
17. **Maricle, D. L. and Hodgson, W. G.**, Reduction of oxygen to superoxide anion in aprotic solvents, *Anal. Chem.*, 37, 1562, 1965.
18. **Valentine, J. S. and Curtis, A. B.**, A convenient preparation of solutions of superoxide anion and the reaction of superoxide anion with copper(II) complexes, *J. Am. Chem. Soc.*, 97, 224, 1975.
19. **Behar, D., Czapski, G., Rabani, J., Dorfman, L. M., and Schwarz, H. A.**, The acid dissociation constant and decay kinetics of the perhydroxyl radical, *J. Phys. Chem.*, 74, 3209, 1972.
20. **Fielden, E. M., Roberts, P. B., Bray, R. C., Lowe, D. J., Mautner, G. N., Rotilio, G., and Calabrese, L.**, The mechanism of action of superoxide dismutase from pulse radiolysis and electron paramagnetic resonance, *Biochem. J.*, 139, 49, 1974.
21. **Holroyd, R. A. and Bielski, B. H. J.**, Photochemical generation of superoxide radicals in aqueous solutions, *J. Am. Chem. Soc.*, 100, 5796, 1978.
22. **Peover, M. E. and White, B. S.**, Electrolytic reduction of oxygen in aprotic solvents; the superoxide ion, *Electrochim. Acta*, 11, 1061, 1966.
23. **Kastening, B. and Kazemitard, G.**, Electrochemische Reduktion von Sauerstoff zum Superoxid-Anion in wasseriger Lösung, *Ber. Bunsenges. Phys. Chem.*, 74, 557, 1970.
24. **Chevalet, J., Rouelle, F., Gierst, L., and Lambert, J. P.**, Electrogeneration and some properties of the superoxide ion in aqueous solutions, *J. Electroanal. Chem.*, 39, 201, 1972.
25. **Divišek, J. and Kastening, B.**, Electrochemical generation and reactivity of the superoxide ion in aqueous solutions, *J. Electroanal. Chem.*, 65, 603, 1975.
26. **McClune, G. J. and Fee, J. A.**, A simple system for mixing miscible organic solvents with water in 10-20 ms for the study of superoxide chemistry by stopped-flow methods, *Biophys. J.*, 221, 65, 1978.
27. **Samuni, A. and Czapski, G.**, Complexes of peroxy radical with transition metal ions, *J. Phys. Chem.*, 74, 4592, 1970.
28. **Knowles, P. F., Gibson, J. F., Pick, F. M., and Bray, R. C.**, Electron-spin-resonance evidence for enzymic reduction of oxygen to a free radical, the superoxide ion, *Biochem. J.*, 111, 53, 1969.
29. **Ono, Y., Matsumura, T., Kitajima, N., and Fukuzumi, S.**, Formation of superoxide ion during the decomposition of hydrogen peroxide on supported metals, *J. Phys. Chem.*, 81, 1307, 1977.
30. **Ballou, D., Palmer, G., and Massey, V.**, Direct demonstration of superoxide anion production during the oxidation of reduced flavin and of its catalytic decomposition by erythrocuprein, *Biochem. Biophys. Res. Commun.*, 36, 898, 1969.
31. **Cohen, G. and Heikkila, R. E.**, The generation of hydrogen peroxide, superoxide radical, and hydroxyl radical by 6-hydroxydopamine, dialuric acid and related cytotoxic agents, *J. Biol. Chem.*, 249, 2447, 1974.
32. **Symonyan, M. A. and Nalbandyan, R. M.**, Generation of superoxide radicals in alkaline solutions of hydrogen peroxide and the effect of superoxide dismutase on this system, *Biochim. Biophys. Acta*, 583, 279, 1979.
33. **Erdey, L. and Inczedy, J.**, Kinetic data on the decomposition of hydrogen peroxide in an alkaline medium, *Acta Chim. Acad. Sci. Hung.*, 7, 93, 1955.
34. **Dorabialska, A. and Kolodziejczak, K.**, Microcalorimetric examination of H_2O_2 decomposition in alkaline solutions, *Zesz. Nauk. Politech. Lodz. Chem.*, 8, 3, 1960.
35. **Fridovich, I.**, Quantitative aspects of the production of superoxide anion radical by milk xanthine oxidase, *J. Biol. Chem.*, 245, 4053, 1970.
36. **McCord, J. M., Beauchamp, C. O., Goscin, S., Misra, H. P., and Fridovich, I.**, Superoxide and superoxide dismutase in *Oxidases and Related Redox Systems*, Vol. 1, King, T. E., Mason, S., and Morrison, M., Eds., University Press, Baltimore, 1973, 51.
37. **Olson, J. S., Ballou, D. P., Palmer, G., and Massey, V.**, The reaction of xanthine oxidase with molecular oxygen, *J. Biol. Chem.*, 249, 4350, 1974.
38. **Olson, J. S., Ballou, D. P., Palmer, G., and Massey, V.**, The mechanism of action of xanthine oxidase, *J. Biol. Chem.*, 249, 4363, 1974.
39. **Fridovich, I. and Handler, P.**, Detection of free radicals generated during enzymatic oxidations by the initiation of sulfite oxidation, *J. Biol. Chem.*, 236, 1836, 1961.
40. **Greenlee, L., Fridovich, I., and Handler, P.**, Chemiluminescence induced by operation of iron-flavoproteins, *Biochemistry*, 1, 779, 1962.
41. **Rajagopalan, K. V. and Handler, P.**, Hepatic aldehyde oxidase. II. Differential inhibition of electron transfer to various electron acceptors, *J. Biol. Chem.*, 239, 2022, 1964.
42. **Arneson, R. M.**, Substrate-induced chemiluminescence of xanthine oxidase and aldehyde oxidase, *Arch. Biochem. Biophys.*, 136, 352, 1970.

43. **Miller, R. W. and Massey, V.**, Dihydroorotic dehydrogenase I. Some properties of the enzyme, *J. Biol. Chem.*, 240, 1453, 1965.
44. **Miller, R. W. and Massey, V.**, Dihydroorotic dehydrogenase. II. Oxidation and reduction of cytochrome *c*, *J. Biol. Chem.*, 240, 1466, 1965.
45. **Handler, P., Rajagopolan, K. V., and Aleman, V.**, Structure and function of iron-flavoproteins, *Fed. Proc.*, 23, 30, 1964.
46. **Nakamura, S.**, Initiation of sulfite oxidation by spinach ferredoxin-NADP reductase and ferredoxin system: a model experiment on the superoxide anion radical production by metalloflavoproteins, *Biochem. Biophys. Res. Commun.*, 41, 177, 1970.
47. **Nakamura, S.**, Stimulation of oxygen uptake of ferredoxin-NADP-reductase-ferredoxin complex by cytochrome *c*, *Biochem. Biophys. Res. Commun.*, 48, 1215, 1972.
48. **Srinivasan, V. S., Podolski, D., Westrick, N. J., and Neckers, D. C.**, Photochemical generation of O_2^- by Rose Bengal and $Ru(bpy)_3^{2+}$, *J. Am. Chem. Soc.*, 100, 6513, 1978.
49. **Anderson, C. P., Salmon, D. J., Mayer, T. J., and Yong, R. C.**, Photochemical generation of $Ru(bpy)_3^+$ and O_2^-, *J. Am. Chem. Soc.*, 99, 1980, 1977.
50. **Harbour, J. R., Chew, V., and Bolton, J. R.**, An ESR study of the spin adducts of OH and HO_2 radicals with nitrones in the ultraviolet photolysis of aqueous hydrogen peroxide solutions, *Can. J. Chem.*, 52, 3549, 1974.
51. **Harbour, J. R. and Hair, M. L.**, Superoxide generation in the photolysis of aqueous cadmium sulfide dispersions. Detection by spin trapping, *J. Phys. Chem.*, 81, 1791, 1977.
52. **Whitten, D. G.**, Photoinduced electron-transfer reactions of metal complexes in solution, *Acc. Chem. Res.*, 13, 83, 1980.
53. **Eriksen, J., Foote, C. S., and Parker, T. L.**, Photosensitized oxygenation of alkenes and sulfide via a non-singlet-oxygen mechanism, *J. Am. Chem. Soc.*, 99, 6455, 1977.
54. **Rabani, J. and Nielsen, S. O.**, Absorption spectrum and decay kinetics of O_2^- and HO_2 in aqueous solutions by pulse radiolysis, *J. Phys. Chem.*, 73, 3736, 1969.
55. **Bielski, B. H. J.**, Reevaluation of the spectral and kinetic properties of the HO_2 and O_2^- free radicals, *Photochem. Photobiol.*, 28, 645, 1978.
56. **Bielski, B. H. J. and Allen, A. O.**, Mechanism of the disproportionation of superoxide radicals, *J. Phys. Chem.*, 81, 1048, 1977.
57. **Klug-Roth, D., Fridovich, I., and Rabani, J.**, Pulse-radiolytic investigations of superoxide catalyzed disproportionation. Mechanism for bovine superoxide dismutase, *J. Am. Chem. Soc.*, 95, 2786, 1973.
58. **Brigelius, R., Spöttle, R., Bors, W., Lengfelder, E., Saran, M., and Weser, U.**, Superoxide dismutase activity of low molecular weight Cu^{2+}-chelates studied by pulse radiolysis, *FEBS Lett.*, 47, 72, 1974.
59. **Pick, M., Rabani, J., Yost, F., and Fridovich, I.**, The catalytic mechanism of the manganese-containing superoxide dismutase of Escherichia coli studied by pulse radiolysis, *J. Am. Chem. Soc.*, 96, 7329, 1974.
60. **Brigelius, R., Hartmann, H. J., Bors, W., Saran, M., Lengfelder, E., and Weser, U.**, Superoxide dismutase activity of $Cu(Tyr)_2$ and Cu, Co-erythrocuprein, *Z. Physiol. Chem.*, 356, 739, 1975.
61. **Weinstein, J. and Bielski, B. H. J.**, Reaction of superoxide radicals with copper(II)-histidine complexes, *J. Am. Chem. Soc.*, 102, 4916, 1980.
62. **Weser, U., Sellinger, K. H., Lengfelder, E., Werner, W., and Strahle, J.**, Structure of Cu_2(Indomethacin)$_4$ and the reaction with superoxide in aprotic systems, *Biochim. Biophys. Acta*, 631, 232, 1980.
63. **Klug-Roth, D. and Rabani, J.**, Pulse radiolytic studies on reactions of aqueous superoxide radicals with copper(II) complexes, *J. Phys. Chem.*, 80, 588, 1976.
64. **Rabani, J., Klug-Roth, D., and Lilie, J.**, Pulse radiolytic investigation of the catalyzed disproportionation of peroxy radicals. Aqueous cupric ions, *J. Phys. Chem.*, 77, 1169, 1973.
65. **Meisel, D., Levanon, H., and Czapski, G.**, Hydroperoxyl radical reactions. II. Cupric ions in modulated photolysis. Electron paramagnetic resonance experiments, *J. Phys. Chem.*, 78, 779, 1974.
66. **Bielski, B. H. J. and Shiue, G. G.**, Reaction rates of superoxide radicals with the essential amino acids, in *Oxygen Free Radicals and Tissue Damage*, Series 65 (new series), Ciba Foundation, Elsevier/North-Holland, 1979, 52.
67. **Jayson, G. G., Parson, B. J., and Swallow, A. J.**, The mechanism of the Fricke dosimeter, *Int. J. Radiat. Phys. Chem.*, 7, 363, 1975.
68. **Pick-Kaplan, M. and Rabani, J.**, Pulse radiolytic studies of aqueous $Mn(ClO_4)_2$ solutions, *J. Phys. Chem.*, 80, 1840, 1976.
69. **Bielski, B. H. J. and Chan, P.**, Products of reaction of superoxide and hydroxyl radicals with Mn^{++} cation, *J. Am. Chem. Soc.*, 100, 1920, 1978.
70. **Bray, R. C., Mautner, G. N., Fielden, E. M., and Carle, C. I.**, Studies on the superoxide ion: complex formation with barium and calcium ions detected by e.p.r. spectroscopy and kinetically in *Superoxide and Superoxide Dismutases*, Michelson, A. M., McCord, J. M., and Fridovich, I., Eds., Academic Press, London, 1977, 61.

71. **Sellers, R. M. and Simic, M. G.**, Pulse radiolysis study of reactions of some reduced metal ions with molecular oxygen in aqueous solution, *J. Am. Chem. Soc.*, 98, 6145, 1976.
72. **Ilan, Y. A., Czapski, G., and Ardon, M.**, The formation of CrO_2^{2+} in the reaction of Cr^{++} + O_2 in aqueous acid solutions, *Isr. J. Chem.*, 13, 15, 1975.
73. **McClune, G. J., Fee, J. A., McCluskey, G. A., and Groves, J. T.**, Catalysis of superoxide dismutation by iron-ethylenediaminetetraacetic acid complexes. Mechanism of the reaction and evidence for the direct formation of an Iron(III)-EDTA-peroxo complex from reaction of superoxide with Iron(II)-EDTA, *J. Am. Chem. Soc.*, 99, 5220, 1977.
74. **Fee, J. A., Shapiro, E. R., and Moss, T. H.**, Direct evidence for manganese(III) binding to the manganosuperoxide dismutase of Escherichia coli B, *J. Biol. Chem.*, 251, 6157, 1976.
75. **Howie, J. K. and Sawyer, D. T.**, Manganese (II) and manganese (III) 8-quinolinol complexes. Redox model for mitochondrial superoxide dismutase, *J. Am. Chem. Soc.*, 98, 6698, 1976.
76. *Nat. Bur. Standards (U.S.) Techn. Note*, 270-3, 1968.
77. **Airey, P. L. and Sutton, H. C.**, Electrochemistry of the HO_2 and O_2^- radicals under steady state conditions. I, *Farad. Trans. I*, 72, 2441, 1976; II, *Farad. Trans. I*, 72, 2452, 1976.
78. **Ilan, Y., Czapski, G., and Meisel, D.**, The one-electron transfer redox potentials of free radicals, *Biochim. Biophys. Acta*, 430, 209, 1976.
79. **Meisel, D. and Czapski, G.**, *J. Phys. Chem.*, 79, 1503, 1975.
80. **See Discussion in** *Photochem. Photobiol.*, 28, 658, 1978.

APPENDIX

ABBREVIATIONS

ADP	adenosine 5′-diphosphate
ATP	adenosine 5′-triphosphate
BPN	*N*-*tert*-butyl-α-phenylnitrone
BSA	bovine serum albumin
CCP	cytochrome *c* peroxidase
CD	circular dichroism
CDAA	*N,N*-diallyl, 2-choroacetamide
CDMPO	2-carboxy-5,5-dimethyl-1-pyrroline-*N*-oxide
CL	chemiluminescence
CMU	monuron
CNA	9-cyanoanthracene
CuDIPS	Cu(II)(3,5-diisopropylsalicylate)
Cu-Zn SOD	copper and zinc-containing superoxide dismutase
cyclic AMP	3′,5′-cyclic adenylic acid
cyt *c*	cytochrome *c*
DABCO	1,4-diazobicyclo[2.2.2]octane
DCA	9,10-dicyanoanthracene
DCIP	dichlorophenol-indophenol
DCMU	dichlorophenyldimethylurea or diuron
DDC	diethyldithiocarbamate
DETAPAC	diethylenetriaminepentaacetic acid
DMF	dimethylformamide
DMH	1,2-dimethylhydrazine
DMPO	5,5′-dimethyl-1-pyrroline-1-oxide
DMSO	dimethylsulfoxide
DPBF	1,3-diphenylisobenzofuran
E.D.	electron donor
EDTA	ethylenediamine tetraacetate, or tetraacetic acid
EAC	Ehrlich ascites carcinoma
EPR	electron paramagnetic resonance
or	
ESR	electron spin resonance
FAD	flavin adenine dinucleotide
FeSOD	iron-containing superoxide dismutase
FMN	flavin mononucleotide
GC	gas chromatography
GDH	glyceraldehyde-3-phosphate dehydrogenase
GLC	gas liquid chromatography
GSH	reduced glutathione
GSSG	oxidized glutathione, or glutathione disulfide
HPLC	high pressure liquid chromatography
HRP	horseradish peroxidase
JS	Jensen's sarcoma
KMBA	4-keto-mercapto-butyric acid
KO_2	potassium superoxide
LDH	lactate dehydrogenase
MDA	malondialdehyde
4-MeP_4BN	4-(*N*-methyl pyridinion) *tert*-butyl nitrone
MnSOD	manganese-containing superoxide dismutase
$NaBH_4$	sodium borohydride
NAD^+	oxidized nicotinamide adenine dinucleotide
NADH	reduced nicotinamide adenine dinucleotide
$NADP^+$	oxidized nicotinamide adenine dinucleotide phosphate
NADPH	reduced nicotinamide adenine dinucleotide phosphate
NBT	nitroblue tetrazolium
NHE	normal hydrogen electrode
NMR	nuclear magnetic resonance

NtB	2-nitroso-2-methylpropane
$^1O_2$ or $^1\Delta_g$:	singlet oxygen
PBN or BPN	α-phenyl-*tert*-butyl nitrone
PMA	phorbol myristate acetate
PM2 CCC-DNA	PM2 covalently closed circular DNA
PMN	polymorphonuclear leukocytes
p-NDA	*p*-nitrosodimethylaniline
4-POBN	α-4-pyridyl-*N*-oxide *tert*-butyl nitrone
R.B.	Rose bengal
RBC	red blood cell
SOD	superoxide dismutase
SQc	ubisemiquinone associated with ubiquinol-cytochrome *c* reductase
SQn	ubisemiquinone associated with NADH-dehydrogenase
SQs	ubisemiquinone associated with succinate-ubiquinone reductase
SV40	simian virus 40
THF	tetrahydrofuran
Tiron	1,2-dihydroxybenzene-3,5-disulfonate
TMN	tetranitromethane
TMPO	2,5,5-trimethyl-1-pyrroline-*N*-oxide
TPA	12-*O*-tetradecanoyl-phorbol-13-acetate
TSY	trypticase soy yeast extract medium
UQH_2	ubiquinol or ubiquinone

INDEX

A

B

C

D

E

T

U

V

W